PROVING HISTORY

Richard C. Carrier

PROVING HISTORY

BAYES'S THEOREM *and the*
Quest for the HISTORICAL JESUS

Prometheus Books

59 John Glenn Drive
Amherst, New York 14228–2119

Published 2012 by Prometheus Books

Cover image © 2012 Media Bakery
Cover design by Nicole Sommer-Lecht

Inquiries should be addressed to
Prometheus Books
59 John Glenn Drive
Amherst, New York 14228–2119
VOICE: 716–691–0133
FAX: 716–691–0137
WWW.PROMETHEUSBOOKS.COM

16 15 14 13 12   5 4 3 2 1

Library of Congress Cataloging-in-Publication Data

Carrier, Richard, 1969–
   Proving history : Bayes's theorem and the quest for the historical Jesus / by Richard
Carrier.
      p. cm.
   Includes bibliographical references and index.
   ISBN 978–1–61614–559–0 (cloth)
   ISBN 978–1–61614–560–6 (ebook)
   1. Jesus Christ—Historicity. 2. Bayesian statistical decision theory. I. Title.

BT303.2.C365 2012
232.9'08—dc23
                                                                    2011050577

CONTENTS

PREFACE

This book is the first of two volumes examining a daring question: whether there is a case to be made that Jesus never really existed as a historical person. The alternative is that Jesus originated as a mythical character in tales symbolically narrating the salvific acts of a cosmic being who never walked the earth (and probably never really existed at all). Later, according to theory, this myth was mistaken for history (or deliberately repackaged that way) and then embellished over time. The present book does not test that claim (as the next volume, *On the Historicity of Jesus Christ*, will), but rather begins the inquiry by resolving the central problem of method: *How* does one test a claim like that? Indeed, how do we test historical theories of any sort whatever? As a result, this book is of interest to all historians, even those who have no interest in the Jesus question. For here I shall explore and establish the formal logic of all historical argument.

All historians have biases, but sound methods will prevent those from too greatly affecting our essential results. No progress in historical knowledge, in fact, no historical knowledge at all, would be possible without such methods. Hence, the aim here is to develop a formal historical method for approaching this (or any other) debate, which will produce as objectively credible a conclusion as any honest historian can reach. One need merely plug all the evidence into that method to get a result. That's a bold claim, I know; but the purpose of this book is to convince you, and if in the end you are convinced, provide the background necessary to implement the method I propose. All I ask is that you give my argument a fair hearing.

You may still want to know what my biases are. I am a marginally renowned atheist, known across America (and many other corners of the world) as an avid defender of a naturalist worldview and a dedicated opponent of the abuse of history in the service of supernaturalist creeds. I am a historian by training and trade (I received my PhD in ancient history from Columbia University) and a philosopher by experience and practice (I have published peer-reviewed articles in the field and am most widely known for my book on the subject, *Sense & Goodness Without God: A Defense of Metaphysical Naturalism*). I have always assumed without worry that Jesus was just a guy, another merely human founder of an entirely natural religion (whatever embellishments to his cult and story may have followed). I'd be content if I were merely reassured of that fact. For the evidence, even at its best, supports no more startling conclusion. So, I have no vested interest in proving Jesus didn't exist. It makes no difference to me if he did. I suspect he might not have, but then that's a question that requires a rigorous and thorough examination of the evidence before it can be confidently declared. Believers, by contrast, and their apologists in the scholarly community, cannot say the same. For them, if Jesus didn't exist, then their entire worldview topples. The things they believe in (and *need* to believe in) more than anything else in the world will then be under dire threat. It would be hard to expect them ever to overcome this bias, which makes bias a greater problem for them than for me. They *need* Jesus to be real; but I don't need Jesus to be a myth.

Most atheists agree. And yet so much dubious argument has appeared on both sides of this debate, including argument of such a technical and erudite character that laypeople can't decide whom to trust, that a considerable number of atheists approached me with a request to evaluate the arguments on both sides and tell them whose side has the greater merit, or whether we can even decide between them on the scanty evidence we have. That's how my involvement in this matter began, resulting in my mostly (but not solely) positive review of Earl Doherty's *The Jesus Puzzle*. My continued work on the question has now culminated in over forty philanthropists (some of them Christians) donating a collective total of $20,000 for Atheists United, a major American educational charity, to support my research and writing of a series of books, in the hopes of giving both laypeople *and* experts a serious evaluation of the evidence they can use to decide who is more probably right. The first step in that process is to assess the methods so far employed on the subject and replace them if faulty.

Though this is a work of careful scholarship, the nature of its aims and funding necessitate a style that is approachable to both experts and laypeople. By the requirements of my grant, I am writing as much for my benefactors as my fellow scholars. But there is a more fundamental reason for my frequent use of contractions, slang, verbs in the first person, and other supposed taboos: it's how I believe historians should speak and write. Historians have an obligation to reach wider audiences with a style more attractive and intelligible to ordinary people. And they can do so without sacrificing rigor or accuracy. Indeed, more so than any other science, history can be written in ordinary language without excessive reliance on specialized vocabulary (though we do need some), and without need of any stuffy protocols of language that don't serve a legitimate purpose. As long as what we write is grammatically correct, accurate and clear, and conforms to spoken English, it should satisfy all the aims of history: to educate and inform and advance the field of knowledge. This very book has been written to exemplify and hopefully prove that point.

The support I received for this work has been so generous, I must thank Atheists United for all their aid and assistance, and all those individual donors who gave so much, and for little in return but an honest report. No one (not even Atheists United, who provided me with the financial grant in aid, nor any donor to that fund) was given any power to edit or censor the content of this work or to compel any particular result. They all gave me complete academic freedom. That also means I alone am responsible for everything I write. Atheists United wanted to see what I came up with, and trusted me to do good work on the strength of my reputation and qualifications, but they do not necessarily agree with or endorse anything I say or argue. The same follows for any individual donors.

And more particular thanks are in order for them, who made this work possible. Benefactors came not only from all over the United States, but from all over the world—from Australia and Hong Kong to Norway and the Netherlands, even Poland and France. Not all wanted to be thanked by name, but of those who did (or didn't object), my greatest gratitude goes to the most generous contributors: Jeremy A. Christian, Paul Doland, Dr. Evan Fales, Brian Flemming, Scott and Kate Jensen, Fab Lischka, and most generous of all, Michelle Rhea and Maciek Kolodziejczyk. Next in line are those who also gave very generously, including Aaron Adair, John and Susan Baker, Robert A. Bosak Jr., Jon Cortelyou, Valerie Mills Daly, Brian Dewhirst, Karim Ghantous, Frank O. Glomset, Paul Hatchman, Jim Lippard, Ryan Miller, Dr. Edwin Neumann, Lillian Paynter, Benjamin Schuldt, Vern Sheppard, Chris Stoffel, James Tracy, Stuart Turner, Keith Werner, Jonathan Whitmore, Dr. Alexander D. Young, Frank Zindler, and Demian Zoer. But I am grateful even for the small donors, whose gifts collectively added up to quite a lot, including the generosity of David F. Browning, David Empey, Landon Hedrick, Gordon McCormick, and many others.

This book would never have been written without all their support. It's very rewarding to present this book to those who respect and enjoy my work enough to keep me employed just to educate themselves and the public about what many consider an obscure issue, and who, above all, have patiently waited so long for the payoff. Special thanks also go to David Fitzgerald (author of *Nailed: Ten Christian Myths That Show Jesus Never Existed at All*), Earl Doherty (author of *The Jesus Puzzle* and *Jesus: Neither God Nor Man*), and Evan Fales (author of the forthcoming *Reading Sacred Texts: An Anthropological Approach to the Gospel of St. Matthew*), for their particular assistance and perspective. I often don't agree with them, but their work did influence me, even if not always in the direction they may have hoped.

CHAPTER 1
THE PROBLEM

Apart from fundamentalist Christians, all experts agree the Jesus of the Bible is buried in myth and legend.[1] But attempts to ascertain the "real" historical Jesus have ended in confusion and failure. The latest attempt to cobble together a method for teasing out the truth involved developing a set of criteria. But it has since been demonstrated that all those criteria, as well as the whole method of their employment, are fatally flawed. Every expert who has seriously examined the issue has already come to this conclusion. In the words of Gerd Theissen, "There are no reliable criteria for separating authentic from inauthentic Jesus tradition."[2] Stanley Porter agrees.[3] Dale Allison likewise concludes, "these criteria have not led to any uniformity of result, or any more uniformity than would have been the case had we never heard of them," hence "the criteria themselves are seriously defective" and "cannot do what is claimed for them."[4] Even Porter's attempt to develop new criteria has been shot down by unveiling all the same problems.[5] And Porter had to agree.[6] The growing consensus now is that this entire quest for criteria has failed.[7] The entire field of Jesus studies has thus been left without any valid method.

What went wrong? The method of criteria suffers at least three fatal flaws. The first two are failures of individual criteria. Either a given criterion is invalidly applied (e.g., the evidence actually fails to fulfill the criterion, contrary to a scholar's assertion or misapprehension), or the criterion itself is invalid (e.g., the criterion depends upon a rule of inference that is inherently fallacious, contrary to a scholar's intuition), or both. To work, a criterion must be correctly applied and its logical validity established. But meeting the latter requirement always produces such restrictions on meeting the former requirement as to make any criterion largely useless in practice, especially in the study of Jesus, where the evidence is very scarce and problematic. The third fatal flaw lies in the entire methodology. All criteria-based methods suffer this same defect, which I call the 'Threshold Problem': At what point does meeting any number of criteria warrant the conclusion that some detail is probably historical? Is meeting one enough? Or two? Or three? Do all the criteria carry the same weight? Does every instance of meeting the same criterion carry the same weight? And what do we do when there is evidence both for and against the same conclusion? In other words, even if meeting the criteria validly increases the likelihood of some detail being true, when does that likelihood increase to the point of being effectively certain, or at least probable? No discussions of these historicity criteria have made any headway in answering this question. This book will.

THE CONSEQUENCES OF FAILURE

The quest for the historical Jesus has failed spectacularly. Several times. Historians now even count the number of times.[8] With the latest quest (numbered "the third") and its introduction of criteria, the concept of Jesus we're supposed to believe existed is actually getting more confused and uncertain the more scholars study it, rather than the other way around. Progress is supposed to increase knowledge and consensus and sharpen the picture of what happened (or what we don't know), not the reverse. Instead, Jesus scholars continue multiplying contradictory pictures of Jesus, rather than narrowing them down and increasing their clarity—or at least reaching a consensus on the scale and scope of our uncertainty or ignorance. More importantly, the many contradictory versions of Jesus now confidently touted by different Jesus scholars are all so very plausible—yet not all can be true. In fact, as only one can be (and that at most), almost all must be *false*. So the establishment of this kind of "strong plausibility" has been decisively proved *not* to be a reliable indicator of the truth. Yet Jesus scholars keep treating it as if it were. This has left us with a confused mass of disparate opinions, vast libraries of theories and interpretations essentially impossible to keep up with, and no real efforts at improving or criticizing the worst and gathering the best into any sort of coherent, consensus view of what actually happened at the dawn of Christianity, or even during its first two hundred years.[9]

I won't recount the whole history of historical Jesus research here, as that has been done to death already. Indeed, accounts of the many "quests" for the historical Jesus and their failure are legion, each with their own extensive bibliography.[10] Just to pick one out of a hat, Mark Strauss summarizes, in despair, the many Jesuses different scholars have "discovered" in the evidence recently.[11] Jesus the Jewish Cynic Sage.[12] Jesus the Rabbinical Holy Man (or Devoted Pharisee, or Heretical Essene, or any of a dozen other contradictory things).[13] Jesus the Political Revolutionary or Zealot Activist.[14] Jesus the Apocalyptic Prophet.[15] And Jesus the Messianic Pretender (or even, as some still argue, Actual Messiah).[16] And that's not even a complete list. We also have Jesus the Folk Wizard (championed most famously by Morton Smith in *Jesus the Magician*, and most recently by Robert Conner in *Magic in the New Testament*). Jesus the Mystic and "Child of Sophia" (championed by Elisabeth Schussler Fiorenza and John Shelby Spong). Jesus the Nonviolent Social Reformer (championed by Bruce Malina and others). Or even Jesus the Actual Davidic Heir and Founder of a Royal Dynasty (most effectively argued in *The Jesus Dynasty* by James Tabor, who also sees Jesus as a kind of ancient David Koresh, someone who delusionally, and suicidally, believed he was sent by God and charismatically gathered followers; not surprising, as Tabor is also a Koresh expert, having been an FBI consultant during the siege at Waco, and subsequently authoring *Why Waco?*). There are even recent versions of Jesus that place him in a different historical place and time, arguing the Gospels were mistaken on when and where Jesus actually lived and taught.[17] Or that conclude astonishing things like that he arranged his own execution to effect a ritual sacrifice to magically cleanse the land.[18] We even get confused attempts to make Jesus everything at once (or half of everything at once, since most theories are too contradictory to reconcile), for instance insisting we should understand him to have been "a prophet in the tradition of Israel's prophetic figures . . . a teacher and rabbi, or subversive pedagogue of the oppressed . . . a traditional healer and exorcist, a shamanistic figure . . . [and] a reputational leader who brokers the justice of Yahweh's covenant and coming reign," whatever that means.[19]

This still isn't even a complete list.[20] As Helmut Koester concluded after his own survey, "The vast variety of interpretations of the historical Jesus that the current quest has proposed is bewildering."[21] James Charlesworth concurs, concluding that "what had been perceived to be a developing consensus in the 1980s has collapsed into a chaos of opinions."[22] The fact that almost no one agrees with anyone else should compel all Jesus scholars to deeply question whether their certainty in their own theory is really even warranted, since everyone else is just as certain, and yet they should all be fully competent to arrive at a sound conclusion from the evidence. Obviously something is fundamentally wrong with the methods of the entire community. Which means you cannot claim to be a part of that community and not accept that there must be something fundamentally wrong with your own methods. Indeed, some critics argue the methods now employed in the field succeed no better than divination by Tarot Card reading—because scholars see whatever they want to see and become totally convinced their interpretation is right, when instead they should see this very fact as a powerful reason to doubt the validity of their methods in the first place.[23]

When everyone picks up the same method, applies it to the same facts, and gets a different result, we can be certain that that method is invalid and should be abandoned. Yet historians in Jesus studies don't abandon the demonstrably failed methods they purport to employ. This has to end. Historians must work together to develop a method that, when applied to the same facts, always gives the same result; a result all historians can agree must be correct (which is to say, the most probable result, as no one imagines certainty is possible, especially in ancient history). If historians can't agree on what that method should be, then their whole enterprise is in crisis, because agreement on the fundamentals of method is the first essential requirement for any community of experts to deem itself an objective profession.

THE SOLUTION

In this book I will present a new method that solves the problems attending the 'method of criteria' so progress can finally be made in the field of Jesus studies. But the method I propose is not limited to that field. It can be employed, and I argue should be employed, in every field of historical study. The quest for the historical Jesus is the principle example on which this book's argument will focus, and this can be taken by Jesus scholars as of direct relevance to their work, but by all other historians as only an example that they can use as a model for adapting the same methodology to any other field or question in history.

The solution I propose involves understanding and applying Bayes's Theorem. To make the case for this, I will have to explain and defend this theorem's structure and application (chapter 3), show how all other valid historical methods actually reduce to it (chapter 4), and exemplify how applying it to specialized questions in history can improve results, in particular by using that theorem to show how and why all historicity criteria in the study of Jesus have failed and what it would take for them to succeed (chapter 5). I then take up more technical questions about the applicability and application of Bayes's Theorem (chapter 6). But before embarking, I must set the groundwork for historical reasoning generally (chapter 2), to make sure we're all on the same page.

CHAPTER 2
THE BASICS

Before proceeding to criticize specific methodologies and propose replacements, it's essential to establish the fundamentals. There is an array of underlying methodological assumptions that all historians should agree on, and which should guide any inquiry into the mechanics of historical method. Misunderstanding can result if these are not laid bare from the start. This chapter surveys these assumptions and why we should all share them.

WHY HISTORY REQUIRES EXPERTISE

To laypeople who ask me what history to trust, I always offer three basic rules: (1) don't believe everything you read; (2) always ask for the primary sources of a claim you find incredible; and (3) beware of scholars who make amazing claims about history but who are not experts in the period, or aren't even experienced historians at all. That three-step guideline provides a basic inoculation against most bad history. But professional historians already know this. Indeed, historians (especially historians of antiquity) know why that third rule is so important. There are four stages of analysis we must complete to credibly examine a historical claim:

First is textual analysis. We have to use the methods of textual criticism and paleography to ascertain whether a document we presently have is authentic and accurately reflects its original—since usually only copies of copies exist today. Critical editions of ancient texts are the result of this process, but reading them requires knowledge and skill, which amateurs rarely have (or have enough of). The corollary for physical artifacts and sites (and other nontextual evidence from the past) is the establishment of authenticity and provenance, which is the purview of archaeologists, antiquarians, and forensic specialists, whose function and importance is the same. Doing all this, or assessing whether it has been done well and what its results mean, requires skill, training, and experience.

Second is literary analysis. We have to ascertain what the author of a text meant. And that requires a good understanding of the language as it was spoken and written in that time and place, and a strong grasp of the historical, cultural, political, social, economic, and religious context in which it was written, since all of that would be on the mind of both author and reader, and this would illuminate, motivate, or affect what was written. Conclusions regarding a text's genre and function fall into this category. Yet knowing all that requires considerable experience studying all those aspects of the culture in question, which is a major factor that separates professionals from amateurs. This is especially true because cultural contexts and assumptions have changed considerably, so intuition will often fail anyone who is not well familiar with those changes. What we assume about genre and the semantic ranges of words in an ancient text, for example, if based solely on experience with modern literature and vocabulary, will be wrong; likewise for assumptions about what was normal, known, or believed in that time and place. Ancient literary genres differed from ours. The range of meanings their words conveyed was different. People knew and believed different things and had very different ideas about what was normal or unusual, and indeed different things *were* normal or unusual than are now. It's impossible to know what all these differences were without broad and extensive study.

Third is source analysis. We must try to identify and assess an author's sources of information. This, in turn, requires knowing a lot about what sources existed then and survive now, what sort of sources an author will have used in that time and place (which again requires vast contextual knowledge only professionals tend to have), and, methodologically, how to ascertain when a particular claim or passage uses a source at all, or some known or hypothesized source in particular (skills amateurs might not be so adept at). More often than not, especially in ancient history, we simply will never know what sources an author used or if they even used any, and even when they tell us, we know they often lied or failed to accurately tell us just what it is they got from that source as opposed to assumed or added themselves. But still key is knowing what was available and what was typical, in terms of what kinds of sources existed and how they were used. Thus, in both respects, expertise is indispensable: knowing the overall milieu of ancient methods and behaviors, and knowing the specific authors and "source books" that existed and what they said. Amateurs typically don't have this information.

Fourth, and only last, is historical analysis proper. This cannot proceed until the first three stages have been completed. Those three stages are often completed by other experts and specialists upon whose work subsequent experts rely. We build on each other's work in that respect; for example, trusting that textual critics have produced the most reliable reconstructed text possible of an ancient book (like the Bible). But knowing when to rely on that groundwork, and how to understand it correctly and critically, requires its own set of skills and experience. Moreover, the skills involved in all four levels of analysis, in particular the fourth, require honing under the guidance of watchful experts, since experience only teaches when you know when you've done something wrong, so you can correct and avoid such mistakes in the future. Without such feedback you can never course-correct, and you will simply persist in any errors you became enamored with at the start. You must have experts at hand to guide you for many sustained years, such as professors and professional advisors, who can quickly identify common mistakes in your arguments, or, by informing you of facts overlooked, reveal how easy it is to overlook important facts and show you how to avoid doing that. Then, they must ensure that you draw conclusions from those facts correctly (which means logically). All so you will learn what constitutes an error and how to correct or avoid such errors in future. Professional historians receive years of training like this. Accordingly, their subsequent work product is much more reliable.

For all these reasons, laypeople must depend on qualified experts to report to them what most likely happened in the past and why. Laypeople need only have the skill to discern experts from nonexperts and to discern when to trust what an expert says. This book is not about that, but about what methods experts themselves should employ. But laypeople can improve their discernment of expert testimony if they, too, are familiar with expert methods, at least well enough to know when an expert is actually using them. This chapter will lay the initial groundwork for that by defining the basic principles of expert history that are *not* controversial. The rest of this book will then propose that something new be added to the arsenal of professional historical method.

THE AXIOMS OF HISTORICAL METHOD

Professional historical inquiry should be based on a set of core epistemological assumptions that I call the *axioms of historical method*. I take each as axiomatic, not in the sense that I can't defend them (I certainly can), but in the sense that insofar as they are to be defended, that defense comes from the field of philosophy, not history. If anyone rejects my axioms, then no further dialogue is possible on this issue until there is agreement on the broader logical and philosophical issues they represent. Producing such agreement is not the point of this book, which is only written for those who already accept these axioms (or who at least agree they should). These **twelve axioms** represent the epistemological foundation of rational-empirical history.[1]

> **Axiom 1:** The basic principle of rational-empirical history is that all conclusions must logically follow from the evidence available to all observers.

By "basic principle" I mean *sine qua non* ("without which, not"), a principle without which you cannot have a rational-empirical inquiry. This means (a) private intuition, personal emotions and feelings, inspiration, revelation, or spirit communications cannot be a primary source of evidence and (b) all conclusions argued from the agreed evidence must be logically valid and free of all fallacies.[2]

> **Axiom 2:** The correct procedure in historical argument is to seek a consensus among all qualified experts who agree with the basic principle of rational-empirical history (and who practice what they preach).

By "correct procedure" I mean this is the only truth-finding procedure that performs well enough to trust—which means this is the only procedure that gets us to the truth at a rate significantly better than chance. By "qualified expert" I recognize this standard comes in degrees (e.g., there is a difference between being qualified merely as a historian and being qualified as a specialist in a specific field like ancient Roman history) and that merely being qualified is not a sufficient condition for knowledge (even the most qualified experts remain unaware of findings and developments in their own field, as there is often far too much written even in highly specialized fields for any mortal to have read it all). Hence, generating consensus is a slow process that radiates outward in circles of authority, from specialists to generalists, as an argument is continually advanced and publicized through proper channels (meaning *established* channels, those which experts themselves trust).

Proper historical argument consists of seeking this growth of consensus and entails everything that that requires (diplomatically, rhetorically, and procedurally—hence the purpose of peer review and my recommended twelve rules, soon to follow). This process cannot be bypassed, as specialists in a field are the most qualified to assess an argument in that field, so if *they* cannot be persuaded, no one should be (unless their resistance can be *proven*—not merely assumed—to have other motives than truth-seeking). Conversely, if they *are* persuaded, everyone else has a very compelling reason to agree (unless, again, their *acceptance* can be proven—not merely assumed—to have other motives than truth-seeking). This is the social function and purpose of having such experts and specialists in the first place.[3]

This consensus-seeking results in a dialectic of criticism and revision, which allows errors and gaps to be identified and corrected and arguments and evidence to be trimmed or fortified, until, by the time an argument reaches a wide consensus, you will have an even stronger conclusion than you started with (since the probability of overlooked facts or errors declines with every peer who examines the case), or else you will have discovered your conclusion is incorrect or unwarranted (as you come to realize the criticisms amount to an adequate refutation). Hence, the process of argument must be to make your best case, then ask the community of qualified experts to rebut it, and then respond to their critique, repeating this cycle of exchange until either of two things happens: you must reasonably accept that your arguments, finally revised in light of wide criticism, do not suffice to justify your conclusion after all; or your critics will accept that their rebuttals are insufficient to warrant rejecting your conclusion. The latter results in a revised consensus on which future scholarship can then build. The former results in putting away an interesting but ultimately untenable theory.

This process works precisely to the extent that every expert adheres to the first axiom. Because conclusions that logically follow from public facts are exactly those in which there should be a consensus, and anyone who accepts such conclusions will join that consensus—whereas conclusions that do not logically follow from public facts *should* be rejected, and will be by anyone devoted to the first axiom. Although they might still accept such conclusions for reasons other than history, by virtue of their commitment to the first axiom they will admit that, and they will distinguish what can be arrived at from historical evidence alone and what can only be arrived at by other means.

It's still true that some experts will give lip service to the first axiom but not follow it, or erroneously believe they are when they aren't. But in the first case they will eventually be exposed by this process as frauds (as more and more peers demonstrate their hypocrisy), and in the second case the process itself will eventually correct them (as more and more peers demonstrate their error). Consequently, this process works the better the more experts in a field adhere to the first axiom competently and consistently. The epistemic reliability of a community of experts can be gauged by exactly that measure.

Axiom 3: Overconfidence is fallacious; admitting ignorance or uncertainty is not.

Ignorance and uncertainty are common and normal. But asserting as known or certain what in fact isn't entails some fallacy in your reasoning. One thing professional historians soon learn is how much we need to accept the fact that we will never know most of what we want to know—such as about Jesus or the origins of Christianity, or anything else in history. Compared to, for example, Richard Nixon or Mark Twain, the documentation for Jesus and the origins of Christianity is *extraordinarily* thin and problematic. And yet even knowing all we'd like to know about Nixon or Twain is impossible, as even for them the evidence is neither complete nor unproblematic; for Jesus and the origins of Christianity, vastly more so.

Anyone who rejects this conclusion is not an objective scholar, but a dogmatist or propagandist whose voice needn't be heeded by any respectable academic community. Likewise, most of what we can say, especially about ancient history, is "maybe" or "probably"—not "definitely." There is obviously more than one degree of certainty. Some things we are more sure of than others, and some things we are only barely sure of at all. Hence, especially in history, and even more so in ancient history, confidence must often be measured in relative degrees of certainty, and not in black-and-white terms of only "true" and "false." Accordingly, historians must be comfortable with ambiguity, uncertainty, and ignorance, and must critically weigh and examine their own confidence in any conclusion. The difference between a real expert and a poser is often evident in that very distinction.

Axiom 4: Every claim has a nonzero probability of being true or false (unless its being true or false is logically impossible).

Not only must we be prepared for uncertainty; we must accept that it's everywhere, differing only in degree. For anything that's possible could yet be true. By "possible" here I mean a claim that is possible *in any sense at all* (as opposed to a claim that is logically impossible), and by "probability" here I mean "epistemic probability," which is the probability that we are correct when affirming a claim is true. Setting aside for now what this means or how they're related, philosophers have recognized two different kinds of probabilities: *physical* and *epistemic*. A **physical probability** is the probability that an event *x* happened. An **epistemic probability** is the probability that our belief that *x* happened is true. For example, the probability that someone's uncle invented the global positioning system is certainly very small (since only a very few people out of the billions living on earth can honestly make that claim). But the probability that your belief "my uncle invented the global positioning system" is true can still be very high. All it takes is enough evidence. The former is a physical probability, the latter an epistemic one. I will establish the proper relationship between physical and epistemic probabilities in chapter 6. For now, know only that unless the context indicates otherwise, when I speak of probability, I mean epistemic probability—though you will notice that often there seems to be no practical difference.[4]

Epistemic probability is the probability that we are correct in any given belief. And with its converse, we measure the probability of being mistaken. For example, if (given all we know at the moment) a claim has a 25% probability of being true, then if we say that claim is true, there is a 75% chance we are mistaken, but if we instead say that claim is *false*, then there is only a 25% chance we are mistaken. Therefore, if we say such a claim is false, we will more likely be correct. And so we say the claim is false. But it still has some probability of being true. Accordingly, when we say something is "probable," we usually mean it has an epistemic probability much greater than 50%, and if we say it's "improbable," we usually mean it has an epistemic probability much less than 50%. Everything else we consider more or less uncertain.

All claims have a nonzero epistemic probability of being true, no matter how absurd they may be (unless they're logically impossible or unintelligible), because we can always be wrong about anything. And that entails there is always a nonzero probability that we are wrong, no matter how small that probability is. And therefore there is always a converse of that probability, which is the probability that we are right (or would be

right) to believe that claim. This holds even for many claims that are supposedly certain, such as the conclusions of logical or mathematical proofs.
For there is always a nonzero probability that there is an error in that proof
that we missed. Even if a thousand experts check the proof, there is still
a nonzero probability that they all missed the same error. The probability
of this is vanishingly small, but still never zero.[5] Likewise, there is always
a nonzero probability that we ourselves are mistaken about what those
thousand experts concluded. And so on. The only exception would be
immediate experiences that at their most basic level are undeniable (e.g.,
that you see words in front of you at this very moment, or that "Caesar
was immortal and Brutus killed him" is logically impossible). But no substantial claim about history can ever be that basic. History is in the past
and thus never in our immediate experience. And knowing what logically
could or couldn't have happened is not even close to knowing what did.
Therefore, all empirical claims about history, no matter how certain, have
a nonzero probability of being false, and no matter how absurd, have a
nonzero probability of being true.

Therefore, because we only have finite knowledge and are not infallible, apart from obviously undeniable things, some probability always
remains that we are mistaken or misinformed or misled. Our challenge,
then, is to believe only claims that we are very unlikely to be mistaken,
misinformed, or misled about. But many things have different levels of
certainty and thus different degrees of probability. And although the probability that a given claim is true (or false) may be vanishingly small and
thus *practically* zero, it is never *actually* zero. It's vital to admit this. For
the truth is not always what is physically most probable, since improbable
things happen all the time. If we know nothing else, often we can still at
least say what's most likely to have happened, and that may then be what's
most credible to believe. But that's *not* the same as saying the alternative
can't be true. We may have to admit it *could* be true, even if we don't think
it is. And we may have to decide just how likely or unlikely either *conclusion* is, quite apart from how likely or unlikely the proposed *event* is.

Almost everything that happens is in some sense improbable—from
the specific conjunction of your own unique DNA to the specific sequence

of people you might meet on any given day. And yet it happens. Though being struck by lightning is very improbable, it nevertheless happens to hundreds of people every year. And if your wallet turns up missing, regardless of which is more probable—it being stolen or your having misplaced it—*either* could turn out to be true. Arriving at a reasonable conclusion as to what is the more likely explanation of any conjunction of facts will require comparing the relative probabilities of all the pertinent evidence on different theories (as I'll demonstrate in subsequent chapters), which requires admitting that theories you don't believe in nevertheless have some probability of being true, and theories you're sure are true nevertheless have some probability of being false. And you have to take seriously the effort to measure those probabilities. For when you do, you may find you can't sustain the level of certainty you once had.

In short, "that's impossible" almost always means, rather, "that's very, very improbable." Once you acknowledge that, you will be forced to ask: How improbable? Too improbable to believe? Why? And how do you know? Any sound methodology must provide the means to answer those questions. Failing to do this is to replace knowledge with thoughtless assumption. For this and all the reasons above, throughout this book whenever I refer to "knowledge" or to what you or we "know," I will not be assuming the philosopher's definition of knowledge ("justified true belief"), but using the terms "knowledge" and "know" as shorthand for "what we think we know, and with very good reason" (in other words, well-justified belief), so that anything we claim to "know" we could be wrong about, but are very unlikely to be.[6] I am therefore excluding mere beliefs (things we aren't sure we know but believe anyway, and things we think we know but without a very good reason) and everything we don't in any sense yet know. Defined this way, knowledge does not require an impossible standard of certainty. It's simply what we are very probably right about.

Axiom 5: Any argument relying on the inference "possibly, therefore probably" is fallacious.

Though we must admit anything that's possible could yet be true, that does not argue for anything actually *being* true. This is a form of modal fallacy I call *possibiliter ergo probabiliter* ("possibly, therefore probably") and it's so common in historical argument that it deserves particular attention. Just because you can conceive of a possible alternative explanation does not entail that your alternative is actually more likely (or in any way likely at all). For example, historians will often dismiss an Argument from Silence by proposing some explanation for why a document is silent on that detail. Of course, knowing why we don't have certain evidence still does not change the fact that we don't have that evidence. All you can then say is that this lack of evidence is inconclusive, not that it supports one conclusion over another. But more importantly, just because you can think of a reason to explain away a document's silence does not mean that reason is *probable*, and if it isn't probable, it isn't a valid objection to an Argument from Silence—so treating it as if it were is a fallacy (I'll further discuss the logic of an Argument from Silence in chapter 4, page 117). Likewise, historians often do little more than conceive of some possible 'just so' story to explain the evidence, and assume that because their interpretation fits, therefore it's true. But that's the same fallacy. An infinite array of possible explanations can be deployed for any set of evidence, and quite a lot of them will even seem an uncanny fit, yet at most only one of them can be true, which means merely finding a seemingly uncanny fit between evidence and theory is a completely unreliable method to employ.

In such ways as these, historians often assume they've won their argument if they can think of any possible explanation contrary to their opponents'. But this is the same modal fallacy again—*unless* the alternative is shown to be not merely *possible*, but *highly probable* (or at least probable enough to carry whatever burden is required of the argument). For example, if a historian has a very good reason to *expect* a document's silence on a particular detail, then that silence cannot argue against anything—though neither does this fact argue *for* anything. Likewise, if a fit between text and interpretation can be demonstrated not merely to *seem* uncanny, but to actually *be* uncanny (for example, by proving such a fit is very improbable unless it was the author's actual intent), then that *does* support that interpretation. By the same token, recognizing this fallacy is what permits us to reject absurd claims even though we grant they still have some tiny probability of being true. Like a thousand mathematics professors missing the same mistake in a formal proof: yes, it can happen, but it's so unlikely there will never be a reason to believe it happened, until we can prove it did. Likewise all the canards of radical skeptics, like "you don't know your house really exists, because you could be a brain in a vat." That's just the same fallacy. Yes, we could be a brain in a vat. But that's extraordinarily improbable. Therefore, when we conclude our house really exists, we're still very probably right.

Just as often overlooked (or downplayed) by historians is the fact that many theories we can posit may indeed be true, yet the evidence may still be insufficient to prove them or to warrant believing them. We thus must be willing to admit what we don't know at all, and what we don't know for certain, and not mistake our not knowing whether a theory is true for our knowing it is false. In other words, just because "possibly, therefore probably" is a fallacy doesn't mean a conclusion produced by this fallacy is *false*, it only means a conclusion is not established by merely being possible. Consider the claim that Caesar shaved on the morning of May 12, 52 BCE, or that Caesar once played dice with a hooker named Maxsuma. We have no evidence of these facts (I just made them up). Yet not believing these things is not the same as believing they are false. Either claim may well be true, even if they were improbable. Indeed, even if they were very *probable* they may yet be false. More to the point, the fact that we can easily explain the silence of extant documents on these claims does not constitute a valid argument for believing them. And yet, not believing those things is not the same as not believing Caesar rode a winged horse or that Caesar once camped on the moon. We can express uncertainty on what days Caesar shaved or what hookers he may have diced with, allowing a great deal to be possible on either point without claiming to know, but we *wouldn't* allow the same latitude for his flying horse or imperial moon base—knowing full well the prior probability of either of the latter is far too low to credit, unlike the prior probability of his shaving on any given day or ever having played dice with a hooker of a particular name.

Hence, though "possibly, therefore probably" is fallacious, there is no inherent evil in speculation, or in the exploration of possibilities. There are a great many things that are true about the ancient world and its people and events that we today have no evidence of at all, and we need to take seriously the range of what could yet have happened though it remains unknown to us.[7] But such speculation must be used sparingly and appropriately. Theories and generalizations that are weakly supported need not be dismissed, if their plausibility can be adequately defended, but they must never be treated as anything more than they are—plausible possibilities, not confirmed facts—unless, of course, the facts at hand make them not merely plausible, but *probable.*

Axiom 6: An effective consensus of qualified experts constitutes meeting an initial burden of evidence.

An effective consensus of qualified experts (by which I mean at least 95 percent agreement) is probably true unless a strong and valid proof arises that it is not. This is a straightforward fact of frequency: the methods that generate such a consensus far more frequently discover the truth than err; therefore, any given result of that consensus is far more probably true than false. And this is a consequence of cumulative probability: it is far more unlikely that an incorrect argument would persuade a hundred experts than that it would persuade only one; and it's far more unlikely that it would persuade any expert than that it would persuade even a hundred amateurs. The fact that the one is harder to do than the other acts as a kind of truth filter: only very well argued conclusions survive it. And that's the very point of requiring it to be survived.

Even if we demonstrated (not merely asserted, but actually proved) that the consensus has been improperly generated (e.g., if we actually proved it is based on dogma or tradition or a repeated error rather than a genuine application of sound methods), that would only be sufficient to establish that the consensus position has not been properly established, *not* that it is false or that any alternative is instead correct. So even under those conditions one must *still* present sufficient evidence for any alternative conclusion in order to *reverse* the consensus. Of course, if properly pursued, such an effort would then aim at *becoming* the consensus. That's how fields advance and make progress; by catching and purging their own errors. But the process must still be followed.

Hence, the burden of proof clearly falls on anyone who would challenge an existing consensus, despite repeated attempts to deny this. For example, in the matter of whether Jesus actually existed as a historical person, historicists have already met the burden of evidence to produce a consensus of qualified experts. So the deniers of historicity must overcome that burden with their own. Attempts to argue that the current consensus has been improperly generated have some merit (e.g., many historicists *do* make assertions out of proportion to the evidence, or simply cite "the consensus" without checking how that consensus was actually generated). But such arguments are still inadequate, since there is certainly *prima facie* evidence for a historical Jesus (hence, historicity need not be asserted dogmatically, even when it is). Moreover, such a claim (that there is an improperly generated consensus) must still meet its *own* burden of evidence. In fact, any claim that the consensus is actually *wrong* (and not merely unfounded) requires meeting an even *greater* burden of evidence (in defense of some alternative theory).

This usually means that a strong consensus of experts entails a high prior probability the consensus theory is true. Yet occasionally this is not so. I believe there is ample reason to conclude that the consensus is not reliable in the study of the historical Jesus and therefore cannot be appealed to as evidence for a conclusion. Yet I still bear the burden of proving that (which I shall partly undertake here in chapter 5, although the facts just surveyed in chapter 1 already go a long way toward making the case). And the *prima facie* evidence for a historical Jesus, which constitutes all the valid evidence the consensus could ever appeal to, still cannot be ignored. But it should be examined anew (a task I'll undertake in the next volume).

Axiom 7: Facts must be distinguished from theories.

Another common error is the conflation of facts with theories. Proper facts are actual tangible artifacts (such as extant manuscripts and archaeological finds) and straightforward generalizations therefrom. Everything else is a theory as to how those facts came about. Some such theories can be so well and widely confirmed that we consider them facts, but one should never assume a theory is that well confirmed before checking to ensure that it is (and merely having an effective consensus of qualified experts is not always enough to confirm this).[8] For example, "the first Christians found Jesus' tomb empty" is not a fact. It is a theory, which is proposed to explain what actually *is* a fact: that certain later stories arose describing such a find, which now survive in various manuscripts. But there are alternative theories of how those facts came about (e.g., that those stories arose originally to convey a symbolic meaning and were embellished later as propaganda).[9] That there are alternative theories of the evidence does not mean those theories are true. It may well be that an actual empty tomb is far more likely. But that has to be argued. It requires sufficient demonstration to warrant belief. It cannot simply be assumed as a fact.

Similarly, if a myth proponent wants to propose that a certain name in the Gospels symbolizes a particular astrological sign, he cannot simply claim that as a fact. It's a theory, and it can only be credible to the degree that it's the best explanation of the facts alleged to prove it (which entails considering what some *other* explanations of that name might be). By conflating facts with theories, very often an entire burden of evidence that is actually required is ignored or never met. This distinction is all the more important for lay readers, who will not know the difference if it's not made clear to them. Consequently, historians should be much clearer than they have been in distinguishing confirmed facts from proposed theories. As a rule, whenever there is a nonnegligible chance you are wrong, you are talking about a theory, not an established fact. And it's often the case that if you have to argue a point, there is a nonnegligible chance you are wrong.

Axiom 8: A conclusion is only as certain as its weakest premise.

It's essential to watch for the weakest link in any argument, because very often a single weak link will render all resulting conclusions just as weak— or, by their accumulation, even weaker. This frequently happens in historical reasoning when qualifiers are snuck in without accounting for their logical consequences. For example, in any argument, analyzed formally, if there is a premise of the form "maybe *x*," then any conclusion depending on that premise cannot be any more certain than "maybe." This follows for *any* qualifying language (like "probably," "possibly," or "perhaps"). For example:

> MINOR PREMISE: Alexander might have wanted to assassinate his father Philip.
>
> MAJOR PREMISE: If Alexander wanted to assassinate his father, then he probably arranged his assassination.
>
> CONCLUSION: Therefore Alexander probably arranged Philip's assassination.

This is a fallacious conclusion, since the minor premise does not say "probably" but "might," which makes it a weaker premise than the major premise requires. But the conclusion cannot be more certain than the weakest premise. Therefore, the only valid conclusion one could produce here would be "Alexander *might* have arranged the assassination of Philip," which is such a trivial conclusion as to be useless (since it's true of practically anyone of the time, and predictive of nothing).

This may seem an obvious point to make, but ignoring it is a frequent error among historians, who often make bold assertions from more hesitant premises, forgetting that somewhere along the line of their argument one of their key supporting points was more speculative than their conclusion would suggest. And it only takes *one* weak premise to render all resulting conclusions equally weak, no matter how strong every other premise may be or how many other strong premises there are. Yet I've seen both sides of an argument take ambiguous evidence like this and derive unambiguous conclusions from it, making this a remarkably common error.

> **Axiom 9:** The strength of any claim is always proportional to the strength of the evidence supporting it.

In line with the axioms above, certainty must be proportional to evidence. The strength of a claim means the likelihood of it being true, which (as already noted) is the likelihood of our being mistaken if we denied it was true. But the strength of the available evidence is measured not just by quantity. In fact, quantity can mean little if the evidence is not independent. For example, having a thousand copies of a letter does not make the claims in that letter any more true, but having ten different witnesses reporting a fact independently of each other *does* (provided nothing else calls their testimony into question). But apart from quantity, strength is also measured by the certainty, credibility, and uniqueness of any potential causal connection between the evidence and what it is claimed to support, or by how securely and abundantly the evidence establishes a broader generalization that makes a particular claim more likely or unlikely. I'll discuss the issue of evidentiary strength in coming chapters. But in every case, our conclusions must be proportionate. We must never assert a claim to be more certain than the evidence warrants.

Axiom 10: Weak claims that contradict strong claims are probably false (and not the other way around).

Any weak claim will by definition have a lower probability of being true than a strong claim. Therefore, if a weak claim contradicts a strong claim, all else being equal, more probably than not the weak claim is false. This does not entail that the strong claim is true. But it does entail that only the strong claim could then be asserted as the most likely of the two, which means a strong claim can never be refuted with a weaker one. All too often, historians will attempt to rebut a well-supported claim by appealing to a claim whose support is much less secure. Such an approach is fallacious and thus, per the first axiom, should be rejected by all expert historians.

Axiom 11: Generalizations must be supported by evidence, and that evidence must consist of more than one example (or of an example that strongly implies a general trend), and once supported, cannot be ignored.

All too often generalizations are declared (such as about what Romans or Jews typically did or thought or said) without any supporting evidence at all—or with only one instance, which is insufficient to demonstrate what was typical, unless the character of that instance strongly supports a conclusion that it is in fact an instance of what was typical. One should also not confuse what was typical with what was possible, not only because of the fifth axiom, but also because there will often be exceptions to any generalization. If you assert a generalization as absolute (i.e., without exception), that carries a far greater burden of evidence than any more ordinary generalization. For every generalization entails its converse, which must be just as defensible on the same evidence: for example, the converse of the generalization "everyone always read aloud" is the generalization "no one ever read silently." If the evidence is insufficient to support the latter, it's insufficient to support the former. By contrast, evidence supporting "everyone *usually* read aloud" is not sufficient to support "no one *ever* read silently." And 'usually' is already a hard thing to prove from just a few examples.[10]

Importantly for this axiom, generalizations are not limited to historical trends but also include all rules of inference. If you apply a general rule of inference in some cases and not others, then when you don't apply the rule you must be able to apply *another* general rule of inference that justifies not applying that rule in that special case. And that *other* general rule must itself be demonstrably exceptionless (i.e., it always applies in every case) or ultimately supported by one that is. Hence, any system of general rules of inference that you construct and employ must be internally consistent, both in theory *and in practice*. This means you cannot arbitrarily apply or fail to apply some rule simply when it suits you. To the contrary, any such negation of a rule must be justified by another valid rule. Hence, this eleventh axiom invalidates 'cherry-picking' and 'special pleading' and other abuses of logic and evidence.

> **Axiom 12:** When one of us cites a scholar, it should only be assumed we agree with what they say that is essential to the point we cite them for.

This is an axiom we should apply to all scholars and authors, but I have a particular interest in asserting it here. Though I may agree with many of the scholars I cite on much else besides what I cite them for, it's still a fallacy to *assume* I do without evidence to that effect. Of course, proving I agree with some additional point by adducing evidence that I do is not assuming I do but arguing I do, which is valid. But beyond that, it will be a fallacy to argue against me by arguing against something said by someone I cite which is not necessary to anything I myself have said.

This kind of 'baggage' fallacy (often deployed as a variety of the textbook fallacy of "poisoning the well") is common enough to warrant particular condemnation. In fact, I see this fallacy committed so regularly, so widely, by accomplished scholars who ought to know better, that I feel the need to call particular attention to it now, in the hopes it will forestall a repeat performance. If you cite a scholar as proving point A, and that same scholar also argues B, but B is not necessary to A, then it is a fallacy for anyone to assume you agree with B, and a fallacy to employ this assumption to argue that if B is not credible then A is not credible. I call this the 'baggage' fallacy because it amounts to saddling an author with all the 'baggage' attached to the scholars he cites or the views he defends, when such attachment is neither entailed nor warranted. Just because I take certain positions or arrive at certain conclusions is no excuse to impute to me all the baggage that is usually supposed to come along with those positions or conclusions.

For example, when I argue a point (such as that distinct elements of Osiris cult can be seen in early Jesus cult), it might be assumed I agree with something else that supposedly goes along with this (such as that *all* elements of Osiris cult were present in early Jesus cult, or that Jesus is merely Osiris under another name, or that Christians just "borrowed" and "revamped" an Egyptian religion). That would be mistaken. It would likewise be mistaken to assume I agree with every other alleged parallel that has in the past been made between Jesus and other pagan gods, or that I agree with every theory as to why or how such parallels came to exist simply because I agree there are some meaningful parallels with pre-Christian gods. The same fallacy also results when I agree with something a particular book said, or cite it as a reference of importance on a specific subject, and then it's assumed I agree with *everything* that book said or its author elsewhere defends.

It might be argued that a scholar who makes a bad argument can't be trusted to ever make a good one, therefore citing the flaws in argument B as counting against argument A "should" be appropriate, but that's still fallacious. You can't know if any particular argument is good or bad until you actually look at it. Even the most stupid, ignorant, and incompetent author can make a good argument from time to time, documenting all the evidence and making logically valid inferences from it. That may be improbable. Which is why it is perfectly right to ignore an untrustworthy source as being a waste of your time. But it is *not* valid to ignore that source when a real expert finds a point they make credible and cites their work establishing it. And as an expert, I only cite sources that argue for a point in a way that is valid and sound (and of use to you to consult), regardless of what that source does otherwise. That's why you must examine my source's arguments before dismissing them, rather than attacking the author who made them.

In short, do not assume what's not in evidence. I might have many disagreements with the scholars I cite on any given point, but on matters not relevant to that point. Hence I often omit these disagreements. But my omitting them should not be mistaken for my having none. Likewise I emphasize what I believe is most defensible and build a case therefrom. But such a procedure leaves out countless details about which countless questions could be asked. Again, my omitting such things should not be mistaken for my not being aware of them. Still, if any of these omitted details undermine my main argument, it is still valid to call attention to that fact. Because my omission of such questions and details may correctly be taken as indicating that I don't believe they undermine or seriously challenge that argument. In other words, as my research in this matter has been extensive, I believe anything I have omitted can be resolved or worked out in ways perfectly consistent with my core argument. But I could be wrong about that. Hence, I always welcome a sound critique from my peers.

Nevertheless, this axiom cannot justify relying on gratuitously bad scholarship. Those who should get little or no mention are scholars who do not employ an adequate method of citation and referencing and consequently make many dubious or false claims or claims incapable of confirmation. Their work is of no use to laypeople (who can't assess which claims are credible or dubious) and of little use to experts (who have to redo all their research anyway before trusting what they say, which negates the point of reading them). Some scholars straddle the line, having insightful things to say, yet I have to fact-check anything they said before relying on it myself—the punishment for which is that I don't cite them if I had to do the work. I just cite the evidence instead.

THE TWELVE RULES OF HISTORICAL METHOD

In addition to the twelve axioms just explained, there are also **twelve rules** I would like to see all historians consistently follow, in order to make their work more credible and worthwhile, and to make progress possible. Though consensus-challengers are more frequently guilty of not following these rules, no one is without sin on this score, and we all fail at them from time to time (myself included). They are the standards by which we seek to correct ourselves and be corrected by our peers. Again, none should be controversial—except, of course, the second part of rule one.

> **Rule 1:** Obey the Twelve Axioms (given above) and Bayes's Theorem (articulated in the remainder of this book). This does not mean you must use Bayes's Theorem in any mathematical sense, only that any historical argument you employ must not violate Bayes's Theorem.

> **Rule 2:** Develop wide expertise in the period, topics, languages, and materials that you intend to blaze any trails in, *or else* base all your assumptions in these areas on the *established* (and properly cited) findings of those who have.

> **Rule 3:** Check all claims against the evidence and scholarship, especially generalizations and assumptions (i.e., don't assume that because you heard or read it somewhere or it just seems plausible that therefore it's likely or true).

Rule 4: Confirm that an argument follows from the original language of a text with as much assurance as from your preferred translation. And confirm that your preferred translation fits the original context (both textual and sociocultural).

Rule 5: Phrase all your claims for optimal truth value. Use all necessary qualifications; avoid hyperbole; do not state as fact what is not fact or as certain what is not certain; always express degrees of certainty or uncertainty when appropriate; acknowledge the difference between a speculation and an assertion; and concede when more research is needed.

Rule 6: Don't conflate weakly supported claims with strongly supported claims, or confuse theories with facts or speculations with theories. Always be explicit in all your writings as to which is which.

Rule 7: Address all relevant and significant evidence *against* what you claim (including any relevant arguments from silence against what you claim).

Rule 8: Take into account problems of chronological development. Everything changed over time, and documents written much later may or may not reflect earlier views or practices, regardless of what they claim. Hence, for example, any argument for influence requires evidence not just of parallels and similarities but of the causal direction of that influence (although this works both ways: just because one source comes later than another does not entail the causal direction runs the same way, as the later source could still be attesting a tradition that predates the earlier source).

Rule 9: Always cite your primary evidence, or cite sources that either cite the relevant primary evidence themselves or cite further sources that collectively do (primary evidence being the earliest surviving evidence in the chain of causation, e.g., a modern or medieval historian citing an ancient historian is not primary evidence if the original text of that ancient historian survives, because then *that* is the primary evidence). In other words, never make controversial assertions without leaving a trail of sources and evidence sufficient to confirm those assertions are true.

Rule 10: Avoid reliance on scholarship published prior to 1950 and rely as much as possible on scholarship published after 1970. Work published prior to 1950 need not be ignored, but should not be relied upon if at all possible. Except perhaps for archaeology and philology (e.g., observational reporting and textual criticism), old work should be avoided altogether or employed only when supported by later work (or your own independent verification).[11]

Rule 11: Always report what the most recent general scholarship says on a subject, or what the current leading consensus is, if either is different from your own view. Do not give the impression that a view contrary to the leading consensus *is* the consensus or that a maverick view is a normal view.

Rule 12: Admit when you are wrong and publish a correction or revision. Constantly seek expert criticism to refine your work in this very respect.

Adherence to all twelve axioms and all twelve rules should consistently produce reliable history, which will continually improve with constructive debate. The only controversial element I have not defended or explained here is the first rule's insistence on employing a fundamentally Bayesian method in history. So to that I now turn.

CHAPTER 3
INTRODUCING BAYES'S THEOREM

I trust I have managed to reveal one of the undeniably impressive properties of Bayesianism: the more it is attacked, the stronger it gets, and the more interesting the objection, the more interesting the doctrine becomes. This feature, together with the positive successes of Bayesianism and the failures of alternative views, certainly justify giving Bayesianism pride of place among approaches to confirmation and scientific inference.

— John Earman[1]

WHEN DID THE SUN GO OUT?

As the tale is told, on the very day Jesus was crucified (though the Gospels don't agree on what day or year that was), "the sun was eclipsed" (Luke 23:45) and "there was darkness over the whole world from the sixth hour to the ninth" (Mark 15:33, Matthew 27:45, Luke 23:44). Though one might want to rescue the text by claiming they really meant an ordinary cloud-front just happened to blow in over Jerusalem, that's certainly not what these authors meant. They meant a supernatural darkness covered the whole inhabited world—Luke claimed it was an eclipse of the sun. But a three-hour solar eclipse is scientifically impossible, especially near Passover (as all the Gospels claim it was), which was always celebrated during the full moon—when, rather obviously, the moon is on the other side of the planet from the sun and can hardly get in front of it, much less stay there for three whole hours. A real eclipse lasts only minutes in any one place, as the moon moves pretty fast, and is quite far away. It's also impossible for an eclipse to darken the whole earth—because the moon is so small and so distant, it only barely covers the sun at all, plus the earth spins on its axis at a brisk pace of nearly a thousand miles per hour. So a solar eclipse only darkens a long thin track across the earth, and most of that only partially. You have to be directly under a total eclipse to view it, while partial eclipses don't "darken the land."

So there can be no doubt; blotting out the sun over the whole known world for three straight hours would be a paranormal event of the highest order. Even arguable scientific explanations (such as a vast dense cloud of space-dust swiftly drifting through the plane of the solar system between the earth and the sun over just those three hours) would require events so astronomically rare that one might argue its coincidence with the death of a man claiming to be the savior of the world would hardly be credible without concluding some superhuman plan was at work. Of course, perhaps it was this very coincidence that caused this claim to be attributed to him rather than someone else, such that had the sun gone out just one week earlier we'd be talking about the Christian Savior James T. Christ, rebel apocalyptic prophet from Joppa (or whoever you care to imagine), simply because *he* was the one executed at that fortuitous moment. But even then we'd have an astronomical event of the greatest importance that needed our attentive study. This would not be some trivial historical curiosity to file away and forget.

But did it happen at all? It's certainly not likely, since events like that don't just "happen." If they did, we'd see more of them. If Jesus were struck by lightning on the cross, that would be unlikely, too, since that rarely happens to anyone, but it's still within the realm of the naturally credible, since hundreds of people are un-portentously struck by lightning in any given year, and lifting them up on a stick is practically asking for it. But the sun going out for three hours? When has that ever happened? How likely is it ever to happen? We're talking about some pretty long odds here. If someone came to you today and said the sun had gone out for three hours one odd Friday back in 1983, you'd need some pretty darned solid proof before believing them, precisely because such an event is so unprecedented, while human fibbing is not (much less delusion or error). But suppose you checked and found that, indeed, the event was widely documented in newspapers, scientific reports, video recordings, and the memories and memoirs of countless witnesses the world over, with no witness giving any contrary account. You would rightly conclude that the probability that *all* of this evidence was the product of a massive worldwide conspiracy (much less a mass delusion on an unbelievable scale) is surely much lower than the event occurring, however rare and whatever its cause.

That would be easily done in this modern day and age. We'd be ideally positioned with access to living witnesses and documentation of every source and kind. But what if the event had occurred in the first century CE? Though by comparison, our access to the evidence would be greatly impaired (the witnesses are all long dead, for example, and newspapers and video cameras didn't exist), it would not be crippled. The entire world at the time had its astronomers—not only hundreds of them throughout the Roman Empire, spanning the whole West from Britain to Syria and from Germany to Africa, but in China, India, Persia, and Babylon as well, and to a lesser extent even among the ancient civilizations of the Americas. They would not fail to document and discuss the phenomenon, some even in scientifically precise detail. The Roman world was also a highly literary age; hardly any member of the elite wasn't writing books or memoirs. And thousands of scraps of original letters and documents survive in parchments, papyri, and other media (mainly from the sands of Roman Egypt), plus thousands of inscriptions and images carved in stone and other materials (across an even wider area). Some astronomical inscriptions also survive from the ancient Americas, as do histories and documents from ancient China, tablets and records from Babylon, and ancient texts from Persia and India. There could not fail to have been mention or discussion of such a remarkable and terrifying event across many of these cultures among their surviving textual traditions and materials (only the farthest would have missed out, being on the other side of the earth at the time). And if indeed that were the case, we would surely have adequate warrant to believe the sun was blotted out for three hours on the corroborated day— for the probability of the entire world, even cultures thousands of miles apart, involving the entire body of ancient scientists and observers, all suffering a simultaneous mass delusion, or conspiring to doctor the record (or any modern deviant attempting this, much less succeeding at it), is surely much lower than the event happening after all.

On the other hand, the universal *silence* of all these materials, except a single claim in a single religion repeated only in its own documents (and documents relying on those), is extraordinarily improbable—unless the event was entirely made up. Indeed, Christians couldn't even coordinate their own mythology—the Gospel according to John exhibits no awareness of any darkness occurring at Christ's death (see John 19). Only the synoptic Gospels mention it (and they all derive from a single Gospel, Mark), as well as authors using those Gospels. We hear of it nowhere else. So it's more than reasonable to say the sun was not blotted out for three hours. Christians just made that up. And when they did, quite clearly no one was around anymore who cared enough to refute them (or their refutations weren't preserved in the extant record).

The sole alleged exception to this fact even proves the rule. Contrary to the claims of Christian apologists, a lost chronicle of Thallus (in which he supposedly tried to explain this darkness as an ordinary eclipse) can't be dated any more accurately than a century after the fact (so we don't know Thallus wasn't just reading the Gospels and conjecturing a reply), neither can we confirm he ever actually mentioned the event in connection with Jesus (in fact, the evidence suggests he probably didn't), nor that he verified its unusual duration (or even its occurrence). And as far as we can tell, he was neither a witness nor an astronomer, and his text is not only alone, but was not even deemed worth the bother of preserving; hence, actually confirming our conclusion.[2] No one else noticed the event, because it didn't happen. We can be quite certain of this, so improbable is the remaining silence in the evidence—or in other words, so strong is our expectation that the evidence would be quite different indeed, had the event really occurred.

This is a slam-dunk Argument from Silence, establishing beyond any reasonable doubt the nonhistoricity of this solar event (for the logic of all arguments from silence, see chapter 4, page 117). This entails, in turn, that the Gospels, even from the very beginning, contain wildly unbelievable claims of inordinately public events that in fact never occurred, yet were never gainsaid by any of the millions of witnesses who would surely have known better. I'll consider the significance of that fact in my next volume. But here, our focus will be on the logic of the argument.

FROM SCIENCE TO HISTORY

Historians need solid and reliable methods. Their arguments must be logically valid, and factually sound. Otherwise, they're just composing fiction or pseudo-history. Much has been written on the method and logic of historical argument.[3] And yet, though none of it is aware of the fact, all of it could be reduced to a single conclusion: all valid historical reasoning is described by Bayes's Theorem (or BT). What that is, and why it matters, is the subject of this chapter. That it models all valid historical methods will be demonstrated in the next chapter.

In simple terms, Bayes's Theorem is a logical formula that deals with cases of empirical ambiguity, calculating how confident we can be in any particular conclusion, given what we know at the time. The theorem was discovered in the late eighteenth century and has since been formally proved, mathematically and logically, so we now know its conclusions are always necessarily true—if its premises are true. By "premises" here I mean the probabilities we enter into the equation, which are essentially the premises in a logical argument. Since BT is formally valid and its premises (the probabilities we enter into it) constitute all that we can relevantly say about the likelihood of any historical claim being true, it should follow that all valid historical reasoning is described by Bayes's Theorem (whether historians are aware of this or not). That would mean any historical reasoning that cannot be validly described by Bayes's Theorem is itself invalid (all of which I'll demonstrate in the next chapter). There is no other theorem that can make this claim. But I shall take up the challenge of proving that in the next chapter. If I'm correct, and it is true that BT models what all historians actually do when they think and reason correctly about evidence and explanations, historians would do well to know more about it.

In all empirical sciences, usually the objective is to discover and test different theories against the evidence until we can determine, given all we know, which theory is most likely true. The reason we pursue theories, and not merely gather facts, is that the facts alone tell us little about the world. To infer anything from those facts requires a theory; in particular, a theory of how that evidence came about (and what future evidence might come about in similar conditions). The exact masses, velocities, and accelerations of falling and orbiting objects can be documented, for example, but that's useless information unless we infer from all this data a general pattern—like a universal force of gravitation, its strength and behavior, and what this then predicts about the behavior of projectiles or spaceships, and what it explains about the structure and behavior of the solar system we inhabit. The latter is all theory, however certain we may be that it's true. But not all science is about discovering universal laws or processes or predicting the future. Geology and paleontology, for instance, are largely occupied with determining the past history of life on earth and of the earth itself, just as cosmology is mainly concerned with the past history of the universe as a whole. Yet even then these sciences are making predictions, regarding what *ancient* evidence might be found in the future, and what ancient events and processes caused the evidence we've already found.[4]

For example, we can document our testimony to seeing highly compressed rock on a mountaintop with extinct seashells embedded within it. But this information is only useful to us if we can infer from such observations (and others like it) that that rock used to be under the sea and thus has moved from where it once was, and that this rock has been under vast pressures over a great duration after those shells were deposited in it; which are theories of how the evidence came about that, when fleshed out, can predict not only what other discoveries are likely (or unlikely) to be made and where (e.g., other mountains with similar histories will have similar finds waiting for us), but also what's going to happen to regions now beneath the sea millions of years hence. A particular pattern and sequence of layers in a rock formation can even confirm to us specific historical facts, such as exactly when a volcano erupted, a valley flooded, or a meteorite struck the earth thousands of miles away. Such data can tell us where the Mississippi

River used to flow millions of years ago (which was not where it is now), how large it then was, its shape, the plants and animals that lived in and around it (many of which no longer do), and much else. All of these conclusions are theories: theories of how all the evidence came about that survives for us to see today. And these theories also entail predictions of what sorts of things will happen to that river over the next few million years. Those predictions will not be exact (they won't tell us *exactly* where the river will be or what its size or shape will then be or exactly what new plants and animals might inhabit it), but they will be generic (they will tell us what *kinds* of outcomes are possible or impossible, likely or unlikely, in all these respects).

History is the same. The historian looks at all the evidence that exists now and asks what could have brought that evidence into existence. And tautologically speaking, what most likely brought it about is what most likely happened. She can then infer what other evidence could be found someday (whether finding it is at all likely or not), and what couldn't, *if* her theory is true. Just as a geologist can predict where the ancient course of the Mississippi River will likely be confirmed to be if further excavation were possible, so a historian can predict what sorts of documentation could someday be found, and if found what we can expect it to contain, if her theories are true—predicting, again, not in exact details, but in that same generic sense, regarding what kinds of evidence should be expected, if any turns up (which is precisely how historians can do any research at all, knowing what to look for, and hope for, in their inquiry). And just as a geologist can make valid predictions about the future of the Mississippi River, so a historian can make valid (but still general) predictions about the future course of history, if the same relevant conditions are repeated (such prediction will be statistical, of course, and thus more akin to prediction in the sciences of meteorology and seismology, but such inexact predictions are still much better than random guessing). Hence, historical explanations of evidence and events are directly equivalent to scientific theories, and as such are testable against the evidence, precisely because they make predictions about that evidence.[5]

In truth, science is actually subordinate to history, as it relies on historical documents and testimony for most of its conclusions (especially historical records of past experiments, observations, and data). Yet, at the same time, history relies on scientific findings to interpret historical evidence and events. Science and history are thus inseparable. But the logic of their respective methods is also the same. The fact that historical theories rest on far weaker evidence relative to scientific theories, and as a result achieve far lower degrees of certainty, is a difference only in degree, not in kind. Historical theories otherwise operate the same way as scientific theories, inferring predictions from empirical evidence—both actual predictions as well as hypothetical. Because actual predictions (such as that the content of Julius Caesar's *Civil War* represents Caesar's own personal efforts at political propaganda) and hypothetical predictions (such as that if we discover in the future any lost writings from the age of Julius Caesar, they will confirm or corroborate our predictions about how the content of the *Civil War* came about) both follow from historical theories. This is disguised by the fact that these are more commonly called 'explanations.' But theories are what they are.

Theories in history are of two basic kinds: theories of evidence (e.g., how the content of the *Civil War* came to exist and survive to the present day), and theories of events (e.g., how that war got started, why Caesar did what he did, why he won, etc.). In other words, historians seek to determine two things: *what* happened in the past, and *why*. The more scientifically they do this, the better. And that means the more they attend to the logic of their own arguments, their formal validity and soundness, the better. Historians rarely realize the fact, but all sound history requires answering three difficult questions about any particular theory of evidence or events: (1) If our theory is false, how would we know it? (e.g., what evidence might there then be or should there be?) (2) What's the difference between an accidental agreement of the evidence with our theory, and an agreement produced by our theory actually being true—and how do we tell the two apart? (3) How do we distinguish merely plausible theories from provable ones, or strongly proven theories from weakly proven ones? In other words, when is the evidence clear or abundant enough to warrant believing our theory is actually true, and not just one possibility among many? As in natural science, so in history—I believe Bayes's Theorem is the only valid description of how to correctly answer these questions.[6]

WHAT IS BAYES'S THEOREM?

The literature on Bayes's Theorem is vast, and usually technical to the point of unintelligibility for historians. But Eliezer Yudkowsky's web tutorial "An Intuitive Explanation of Bayes' Theorem (Bayes' Theorem for the Curious and Bewildered: An Excruciatingly Gentle Introduction)" (at http://yudkowsky.net/rational/bayes) provides a good introduction to the theorem, how to use it, and why it's so important. His follow-up article, "A Technical Explanation of Technical Explanation" (at http://yudkowsky .net/rational/technical) is even better, and you will find it very useful in a number of ways, but it requires that you gain familiarity with Bayes's Theorem first. In print, Douglas Hunter, *Political-Military Applications of Bayesian Analysis: Methodological Issues*, makes an even more palatable introduction. Hunter provides an extended example of how to employ Bayesian reasoning to history, while Yudkowsky's focus is the sciences, but Hunter still covers all the basics and is a good place to start. Likewise, though Hunter was a CIA analyst and writes about using Bayes's Theorem to assess political situations, the similarities with historical problems are strong, and his presentation is intelligible to beginners, using a minimum of actual math. Similarly approaching the kind of problems historians deal with (and thus worth looking at by way of example) are applications of Bayesian reasoning in legal theory.[7]

Archaeologists are already making serious efforts to employ Bayesian methods, and in quite sophisticated ways. Though their questions and techniques are more advanced than most historians need, the underlying principles, introductory explanations, and governing logic is often still pertinent to historians in all fields.[8] Wikipedia also provides an excellent article on Bayes's Theorem (though sometimes less trustworthy in other areas, Wikipedia's content in math and science now tends to surpass even print encyclopedias), although in some respects too advanced for laypeople. But if you do want to advance to more technical issues of the application and importance of Bayes's Theorem, there are several highly commendable texts.[9]

Formally, the theorem is represented by this rather daunting equation:

$$P(h|e.b) = \frac{P(h|b) \times P(e|h.b)}{[\, P(h|b) \times P(e|h.b) \,] + [\, P(\sim h|b) \times P(e|\sim h.b) \,]}$$

I will explain the specific terms in this equation later.[10] For now you only need know that P = probability, h = hypothesis, e = evidence, and b = background knowledge, and all of it roughly translates into English as "given all we know so far," then:

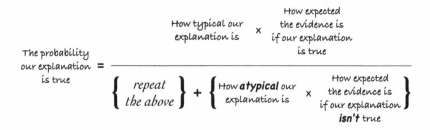

Notice that the bottom expression (the denominator) represents the sum total of all possibilities, and the top expression (the numerator) represents your theory (or whatever theory you are testing the merit of), so we have a standard calculation of odds: your theory in ratio to all theories. The numbers that would go into these terms are probabilities, represented as decimal fractions of 1 (e.g., 25% = 0.25; 80% = 0.80; 100% = 1). Though we don't think in mathematics this way, we are nevertheless doing mathematics intuitively whenever we make any argument for any theory of evidence or events. Every time we say something is "implausible" or "unlikely," for example, we are covertly making a mathematical statement of probability (and if this is not already obvious, I will prove it in the next chapter, beginning on page 110). The fact that we leave the precise figures vague is no excuse not to attend to what those figures could reasonably be (or reasonably *couldn't* be). Because what we actually *mean* when we say things like that has consequences for the logic of any argument we make. Bayes's Theorem simply describes what those consequences are.

The measure of how "typical" our proposed explanation is, is a measure of how often that kind of evidence (or that kind of event) has that kind of explanation (rather than some other). Formally, this is called the **prior probability** (or just "the prior"). I'll discuss this element more later (here and in chapter 6). For now, it's enough to know that whatever probability we assign to this, the term for how *atypical* our explanation is must necessarily equal the converse; that is, if our explanation is "the" explanation in 80% of comparable cases, then its prior probability is 0.80, which in turn means the contrary probability (the measure of its "atypicality") is $1 - 0.80 = 0.20$. The fact that a theory's prior probability is not an absolute probability, but a relative probability, is commonly overlooked yet this is one of the most important features of correct reasoning about a claim's probability. In any kind of causal reasoning, the prior does not measure how often such a thing happens, but how often such a thing happening is the explanation of that kind of evidence (rather than something else explaining that same evidence). For example, if someone claims they were struck by lightning five times in their life, the prior probability they are telling the truth is *not* the probability of being struck by lightning five times, but the probability that someone *in general* who claims such a thing would be telling the truth. In other words, how often such claims are caused by someone actually being struck by lightning five times, relative to how often such claims are caused by error, delusion, or lies. That's the *prior* probability that the claim is true in any specific case, in other words how "typically" such claims turn out to be true. When someone claims they've been struck by lightning five times, that twinge of initial skepticism you feel represents your innate estimate of the prior probability that someone who claims something like that might be lying or mistaken. Of course, the rarity of the event plays a role in that calculation. But it's ultimately the frequency with which claims of such rare events are true that is being measured.[11] And the converse of this probability (that the claim is true) is the *other* prior probability (that the claim is, for whatever reason, false).

But the prior probability alone does not tell us whether a claim is true. We must also consider the evidence available in any specific case. That's the role of the two other terms in the equation, which aren't measuring the prior probability of the hypothesis, but the *likelihood of the evidence*, in other words, how expected that evidence is. The measure of how "expected" the evidence is, is a measure of how likely it is that we would have that evidence (or anything relevantly comparable to it), rather than some other evidence instead, *if* our theory were true. In other words, if our theory is true, then what sort of evidence do we expect, and how well does the evidence we actually have match that expectation? This is measured by a **consequent probability** of the evidence (or just "the consequent").[12] In this respect, 'evidence' includes not just the actual items of evidence we have, but also the evidence we conspicuously *don't* have (despite a reasonably diligent search). For example, if missing evidence is unlikely, yet that expected evidence is missing (as in the example of the disappearing sun with which we began), then the consequent probability is low. This is called an *Argument from Silence*—the validity of which I'll examine in the next chapter (page 117). On the other hand, if the evidence we have is pretty much exactly the sort of evidence we should expect to have, then the consequent probability is high.

The last term in the formula is a similar measure of how expected the evidence is *if our theory is false*. Unlike prior probability, this is not equal to the converse of the other term, for these two consequent probabilities don't have to sum to one.[13] They are measured independently of each other. They can even both equal one. If the evidence is exactly what we should expect regardless of whether the theory we are testing is true or false, then the consequent probabilities are indeed both one, and in such a case we simply don't have any evidence that permits us to tell whether our theory is true or not, apart from its prior probability. For example, if all the evidence we can reasonably expect to have is someone's word, that is exactly the same evidence we would expect to have whether they were lying or telling the truth. Which is why, when we suspect the possibility of lying (i.e., when the prior probability of a lie in that case is not small), we require more evidence than someone's word.

A common mistake is to assume that in estimating this latter consequent probability we are asking how likely the same evidence would be if
nothing were present to cause it. To the contrary, we must ask how likely
that evidence would be if something else *were* present to cause it. In other
words, how likely is the evidence if some *other* explanation is true—some
explanation other than our own. To answer that question, we have to seriously look for, and seriously consider, alternative explanations of the evidence. When there are many possible explanations, the Bayesian formula
can be expanded to account for them all (see page 69), but when only
two of those explanations have any significant likelihood of being true,
then we can treat our theory being false as equivalent to the *alternative*
theory being true. In such a condition other theories would remain logically possible, but far too improbable to credit, so we can safely ignore
them. Because when we ignore such theories, the results we produce using
BT will still be a reasonable enough approximation, just not exact to the
nth decimal place, which is more than adequate for historians, who have no
need of such precision. I'll explain this further in chapter 6 (page 70). For
now, you need only accept that wildly implausible theories can be safely
ignored when estimating the probabilities of more plausible contenders.
And for those, we need estimate only three probabilities: the prior probability a theory is true, the likelihood of the evidence if that theory is true
(in other words, the consequent probability on h), and the likelihood of the
evidence if some other theory is true (the consequent probability on $\sim h$).

When we have reasonable values for all these terms, Bayes's Theorem
entails a particular conclusion as to how probable our theory is—*given
all that we know at that point in time*, since a Bayesian result is a conditional probability. It's conditional on current knowledge, which means if
we discover new theories or facts, the conclusion might change. Bayes's
Theorem thus tells us what we are warranted in believing at any given
time, fully acknowledging that this can change with new information.
If, for example, Bayes's Theorem tells us our theory has a final posterior
probability of 80%, then given what we now know there is an 80% chance
we're right—but that means there is also a 20% chance we're wrong. Since
there will always be some probability our theory is false (however small

that probability may be), this accounts for the possibility that new information could reveal we were wrong all along. But a high probability also entails that such a reversal is unlikely, which is why we are warranted in trusting it. You would say it's then a good epistemic bet.

A BAYESIAN ANALYSIS OF THE DISAPPEARING SUN

Applying all this to the historical claim of a global darkness with which we began this chapter, we can see how the argument presented there in common colloquial terms actually corresponds to the structure of Bayes's Theorem. The theory to be tested (h) is that the sun actually went dark for three hours in the early first century (in other words, that the claim being made is true). The alternative theory ($\sim h$) is that this didn't happen—in other words, that the accounts we have of it were fabricated by storytellers (regardless of who or why or how). The evidence (e) consists of the claims in the Gospels (and sources citing them) *and* the vast and peculiar absence of other evidence outside the Gospels (and sources citing them). Our background knowledge (b), which includes everything we know about science, astronomy, human nature, the society and culture of the first century, and so on, tells us that claims like "the sun was blotted out for three whole hours" (especially when even an ordinary solar eclipse would have been impossible) are almost always caused by someone telling tall tales, and rarely caused by such remarkably rare events actually happening. In other words, countless other cases establish that fibbing or delusion is far more commonly the cause of such tales than actual unprecedented phenomena—the more so in "sacred tales," such as the Gospels.

Though "rarely" certainly doesn't mean "never," the conclusion still follows that the prior probability of h ("the story was told because it was true") is very low, while the prior probability of $\sim h$ ("the story was told because it was made up") is very high, because that's how it usually turns out in such cases. This is, in fact, what we already assume when we are immediately skeptical of wild claims like that (this is sometimes amusingly called the 'Smell Test,' the logic of which I'll examine in the next chapter, page 114). When someone once told me they had seen a demon levitate a girl over her bed for half an hour, I already knew it was very unlikely this story would turn out to be true. That's my intuitive recognition that the prior probability of hallucinating or making that stuff up is far higher than the prior probability of such a claim actually being true. This is because human fibbing, illusion, hallucination, or delusion is far more common, and because such fabulous displays of demonic telekinesis have never been reliably documented, and thus even if they happen, they happen far less often than all those other causes of the same kinds of claims.

Still, the odds aren't zero. There are certainly conceivable (albeit extremely rare) natural explanations of a bizarre solar darkness, so even if I rejected the supernatural outright, I would still have to admit there is some small probability that this darkness really happened. And I can't honestly reject the supernatural outright anyway. For there is always some small probability I'm wrong about there being no supernatural causes or phenomena—even if the odds of my being wrong about that are even lower than the odds of a natural cause of a three-hour blockage of the sun.[14]

But that doesn't conclude the matter. Even claims with a very low prior probability can still turn out to be true—and not only true, but supremely credible—because the consequent probabilities can still diverge enough to overcome even the smallest prior. In other words, when the evidence really is good enough, even the incredibly improbable becomes likely. I can certainly imagine sufficient technical and professional documentation of demon levitations that would convince me demon levitation was real. So could directly witnessing it myself. Indeed, if everything that happened in the movie *Constantine* had actually happened to me, it would be irrational *not* to believe. Such bodies of evidence being fabricated or mistaken (or even hallucinated) would be far less probable than the event simply being true (provided we could confirm, to a high probability, the absence of any likely explanation like drugs or schizophrenia). But a half-hour demon levitation that somehow no one thought even to record on video (despite this being the twenty-first century when even common cell phones can record video) is contrary to reasonable expectation. The absence of evidence here is suspicious, and therefore actually *less* probable if the story is true than if it's false (since the lack of such obvious documentation is always expected for a lie, but not as expected for the real deal).

This would not have been the case twenty years ago, when recording video was expensive and few had the resources for it, such that we would *not* expect video to be made. But now video cameras are everywhere and cost nothing to operate. It would be unusual for someone to know they are observing an incredible event and for a whole half hour not make any effort to record it. It would be many times more unusual if *no one ever* did this when demon levitations are supposed to be frequently occurring *all over the world every year*. And unusual means infrequent, which means improbable. But a liar will always have an excuse for why they didn't do something so obvious. That would not be unusual at all. The difference may be small (perhaps making for a weak Argument from Silence, as I discuss on page 119), but it's still a difference, and it doesn't favor the claim being true. This still doesn't *entail* that lack of documentation makes the claim unlikely. For there can be honest excuses, too. So it merely *lowers* the probability. How much lower will depend on particulars, and the priors. But when the prior is low, even honest excuses cannot make a claim likely. That's precisely why "anecdotal evidence" is worthless in science, and in courts of law. It's not that anecdotal evidence is necessarily false. It's just that it's much too likely to be false. Conversely, the more extensive and reliable the documentation we have, the more a low prior can be overcome. Because the odds of error or fabrication then decline.

Applied to the darkness scenario, the example I gave of a claimed three-hour worldwide darkness in 1983, which we confirmed in all the ways we should expect, demonstrates the same principle: such a scale of evidence is so improbable as a fabrication (or as anything else other than the event actually happening), that even though the story being true has an extremely low prior probability, the evidence in this case would more than overcome it (being *even more* improbable unless the story were true, entailing sufficiently divergent consequent probabilities). In contrast, the vast *absence* of the evidence we should expect from the world's cultures of the first century is vastly *improbable* if *that* story is true, yet entirely expected if it's false. Therefore that claim should not be believed. This kind of demarcation between evidence and background information is characteristic of historical method, in which *e* consists of the evidence to be explained by *h* (and by its competitors, represented by ~*h*), and *b* consists of what has typically happened before, in other relevant cases. Typically suns don't go out for three hours in the middle of the day. That is what we derive from our background knowledge. And that gives us our prior probabilities. The *evidence* in this particular case is then the status of documentation that the event (or its fabrication) is expected to cause. In other words, how typically do we get that kind of evidence, *given that cause* (*h* or ~*h*). And that gives us our consequent probabilities. The equation does the rest.

Representing these two options mathematically, Bayes's Theorem models this very line of argument as follows. I'll start with the hypothetical darkness in 1983. Merely for convenience I will employ the value 0.01 (or 1%) for the prior probability that such a story would be caused by a real unprecedented darkness rather than by being made up. Again, this is *not* the probability of such an unprecedented darkness *occurring*, but the probability that having a story of such a thing would indicate it did. In other words, it's whatever we find to be the typical probability that anyone who told such a story would be telling the truth. I will also use 0.00001 (one in a hundred thousand) for the consequent probability that vast worldwide evidence confirming the event would exist *even if the story were made up* (which is really *far* less likely than one in a hundred thousand). The result would be identical with even vastly smaller numbers than these (like 10^{-9} and 10^{-12}, respectively), since it's their ratio that determines the outcome rather than their actual values—and if anything their ratio would be greater, not smaller, which would confirm my point *a fortiori* (a method I will discuss on page 85). Don't worry too much about the exact details of all the math here. I'll discuss it later. For now just follow along:

h_{1983} = such a darkness happened in 1983

~h_{1983} = the darkness of 1983 is a made-up story

e_{1983} = all the expected documentation is found confirming the darkness in 1983 b   = everything we know about human nature, astrophysics, technology, the culture and society of 1983, etc.

$$P(h_{1983}|e_{1983}.b) = \frac{P(h_{1983}|b) \times P(e_{1983}|h_{1983}.b)}{[\,P(h_{1983}|b) \times P(e_{1983}|h_{1983}.b)\,] \;+\; [\,P(\sim h_{1983}|b) \times P(e_{1983}|\sim h_{1983}.b)\,]}$$

$$= \frac{0.01 \times 1}{[0.01 \times 1] + [0.99 \times 0.00001]}$$

$$= \frac{0.01}{[0.01] + [0.0000099]}$$

$$= \frac{0.01}{0.0100099}$$

$= 0.9990$ (rounded) $= 99.9\% = $ *the probability that this darkness claim* (h_{1983}) *is true, given all the evidence we have and all our current background knowledge.*

So even if there was only a 1% chance such a claim would turn out to be true, that is, a prior probability of merely 0.01, the evidence in this case (e_{1983}) would entail a final probability of *at least* 99.9% that this particular claim is nevertheless true. The prior probability may even be one in a billion, but the consequent probability of the evidence (of e_{1983}) would also more realistically be at least one in a trillion (i.e., still a thousand times less likely), which would produce exactly the same result: a 99.9% chance the claim is true (since the ratio is the same). Thus, even extremely low prior probabilities can be overcome with adequate evidence.

Practically the same result would be obtained for the Gospel claim of a world darkness if we had the evidence we would then expect (vast multicultural attestation). But we don't. Bayes's Theorem models the ensuing argument again, and this time I'm favoring the theory with even more generous numbers:

h_{30s} = such a darkness happened in the 30s CE

$\sim h_{30s}$ = that darkness is a made-up story (deliberately or by hallucination, etc.)

e_{30s} = a collection of interrelated hagiographies composed decades later (i.e., the Gospels) contains the claim (as well as texts using them), but no other documents independently confirm it

b = everything we know about human nature, astrophysics, technology, ancient sacred writings in general, the New Testament documents in particular, early Christianity, the culture and society of Palestine and the Roman Empire in the first century, etc.

$$P(h_{30s} | e_{30s} .b) = \frac{P(h_{30s} | b) \times P(e_{30s} | h_{30s} .b)}{[\, P(h_{30s} | b) \times P(e_{30s} | h_{30s} .b) \,] \;+\; [\, P(\sim h_{30s} | b) \times P(e_{30s} | \sim h_{30s} .b) \,]}$$

$$= \frac{0.01 \times 0.01}{[0.01 \times 0.01] + [0.99 \times 1]}$$

$$= \frac{0.0001}{[0.0001] + [0.99]}$$

$$= \frac{0.0001}{0.9901}$$

$= 0.000101 = 0.01\%$ (rounded) = *the probability that this darkness claim (h_{30s}) is true, given all the evidence we have and all our current background knowledge.*

Here we find the claim is almost certainly false (with odds of around one in ten thousand, and that's *at best*), because the evidence we have is not at all expected on the theory that this three-hour darkness actually happened. Indeed, the odds that we would have such a universal silence of other witnesses is far lower than the one percent I assigned it here. Yet lowering that number reduces the odds of the claim being true *even further* (and as for the previous example, all the same goes for the prior probabilities, too). In contrast, the evidence we have is exactly what we should expect if the story was made up: only three hagiographies, two of them directly derived from the first, and texts relying on these, repeating a mythical claim about a divine hero. Hence, I assigned this evidence a probability of 100 percent (or as near to it as makes no relevant difference mathematically—a distinction I'll say more about in a moment)—because if the story were made up, that's exactly the kind of evidence we'd have.

The point of these examples is to illustrate that how we normally reason about claims like this is exactly described by Bayes's Theorem, even if we never knew that. Bayes's Theorem is thus not an alien way of thinking. It's just an exact model of how we always think (when we think correctly). Thus, when applied correctly, BT will not only represent correct thinking about any empirical claim; it will help us identify and expose incorrect thinking. Because the one thing Bayes's Theorem adds to the mix is an exposure of all our assumptions and how our inferences derive from them. Instead of letting us get away with using vague verbiage about how likely or unlikely things are, Bayes's Theorem forces us to identify exactly what we mean. It thus forces us to confront whether our reasoning is even sound.

WHY BAYES'S THEOREM?

The two main advantages of the Bayesian method are that no one can deny the conclusion who accepts the premises (provided those premises are validly stated within the requirements of the theorem), and it forces us to consider what those premises really ought to be (which is to say, what probabilities we ought to put into the equation), thus pinning down our subjective assumptions and making them explicit and thus accessible to criticism (even by ourselves). Understanding the logical structure of a sound Bayesian argument, as formally represented in BT, can thus prevent historians from making specious or fallacious arguments, or from being seduced by them.

With BT, instead of myopically working out how we can explain all the evidence "with our theory," we start instead by asking how antecedently likely our theory even is, and then we ask how probable all the evidence is on our theory (both the evidence we have, and the evidence we don't) *and* how probable all that evidence would be on some *other* theory (every other theory that has any claim to plausibility, but especially the most plausible alternative). Only then can we work out whether our theory is actually the best one. If we instead just look to see if our theory fits the evidence, we will end up believing any theory we can make fit. And since that will inevitably include dozens of theories that aren't actually true, "seeing what fits" is a recipe for failure. In fact, this is worse than failure, since we will have deceived ourselves into thinking the method worked and our results are correct, because "see how well the evidence fits!" That's the result of failing to take alternative theories of the evidence seriously. That this is exactly what has happened in Jesus studies (as shown in chapter 1) should be proof enough that historians need a new method. One that actually works. And as far as I can see, BT is the only viable contender.

This is all the more important because psychologists have found that this 'see what fits' approach is a slave to *confirmation bias* (where we only see or remember data confirming our hypothesis, and overlook or forget data disconfirming it), which is a fallacious mode of reasoning biologically innate to the human brain (Wikipedia maintains an excellent and well-referenced article on it). Yet it is diametrically opposite the scientific method,[15] which instead tests theories by looking for evidence *against* them (and confirmation then comes only from not finding it), which requires an investigator to imagine what evidence should exist if his theory is *false*.[16] And that requires taking seriously alternative explanations of what a theory is meant to explain. Historians need to do the same.

The applicability of BT to all historical arguments and hypotheses will be proved in the next two chapters. Here I shall only outline the underlying principles and logic that warrant learning and applying BT, and then I'll meet the most common objections to the idea of applying it to history. Why should historians use it? That's the question I'll answer here. Following that, I will address the formal mechanics of BT, and then meet several more technical objections to the idea of applying it to history that then arise.

The first fundamental observation that should open anyone's mind to learning and applying BT is the principle of nonzero probabilities. As discussed in chapter 2, there are two different kinds of probabilities: physical and epistemic. As argued there, the fourth axiom holds: all empirical claims about history, no matter how certain, have a nonzero probability of being false, and no matter how absurd, have a nonzero probability of being true. The only exceptions I noted are claims about our direct uninterpreted experience (which are not historical facts) and the logically necessary and the logically impossible (which are not empirical facts).[17] Everything else has some epistemic probability of being true or false. Once we accept this, Bayes's Theorem applies. Methodologically, for every observed fact, *some* explanation can be devised to explain it away in support of *any* conceivable claim or theory. So you cannot end any debate by declaring that you "can" explain a piece of evidence. Per my fifth axiom (in chapter 2) just because a theory *can* explain a fact doesn't mean that theory *is* the explanation of that fact. The question must be which explanation, among all the viable alternatives, is actually the most likely. And that's where Bayesian reasoning enters in. If all explanations have some probability of being true, the comparison of their probabilities must entail that one of them is more probable than the others—or that none is, in which case we *can't* say which theory is correct. Either way, BT is the only means of sorting this out.

"But what has math to do with history?" The most common objection to this is that BT involves math, and we don't think in math. Historians certainly don't do math. What has mathematics to do with historical reasoning? Bringing numbers into it seems suspect. But that's naive. The reality is that we *do* think in math. All the time. And historians most of all. We just don't know we're doing it. Every time you accept or reject a conclusion because something is "unlikely" or "credible" or "implausible" or "more likely" or "most likely," you are doing math. You're just using ordinary words instead of numbers. Select any claim in the world, and you will immediately be able to say roughly how likely you think it is—in some verbally descriptive way ("very probable," "extremely improbable," "somewhat likely," "as likely as not," etc.). You will even be able to rank many claims in order of their likelihood. And when presented with a new claim, you'll be able to insert it somewhere into that order again where you think it goes. And you do this all the time whenever you sift through competing theories of evidence and events. All of this entails mathematical thinking. Because as soon as you say x is more than y, you are doing math.

In fact, your thinking is even more mathematically precise than that. When you say something is "probably true," you mean it has an epistemic probability greater than 50%. Because that's what that sentence literally means. And when you say something is probably false, you mean it has a probability less than 50%. And when you say you don't have any idea whether a claim is probably true or probably false, you mean it has a probability *of* 50%, because, again, that's what that sentence literally *means*. Likewise, when you say something is "very probably true," you certainly don't mean it has a probability of 51%. Or even 60%. You surely mean better than 67%, since anything that has a 1 in 3 chance of being false is not what you would ever consider "*very* probably true." And if you say something is "almost certainly true," you don't mean 67% or even 90%, but surely at least 99%.

And when you start comparing claims in order of likelihood, you're again thinking numbers. That the earth will continue spinning this summer is vastly more probable than that a local cop will catch a murderer this summer, which is in turn more probable than that it will rain in Los Angeles this summer, which is in turn more probable than that you'll suffer an injury requiring a trip to the hospital this summer. And so on. You certainly don't know what any of these probabilities are. And yet you have *some* idea of what they are, enough to rank them in just this way, and not merely rank them, but also rank them against known probabilities, because you know there is data on the frequency with which people like you get hospitalized for injuries, the frequency with which it rains in L.A., the frequency with which murderers are caught in your county, even the frequency with which the earth keeps spinning every year (we have data on that extending billions of years back, not just for the earth itself, but for all the phenomena that could stop the earth spinning). Thus even a merely ordinal ranking of likelihoods always translates into some range of probabilities. In fact, because you know each is more likely than the next, and roughly how much more likely, probability *ratios* are implicit in your ordinal ranking, and as it happens BT can proceed with just these ratios, without ever knowing any of the actual probabilities (as I show on page 284). And yet you will still often know in what ballpark each probability actually lies, because you can often relate them to a well-quantified benchmark, something whose probability you actually *know*. And when you think about it, you'll agree this knowledge is not completely arbitrary, but entirely reasonable and founded on evidence (such as your own past experience and study of the relevant facts and phenomena). You might never have thought about any of this, but your being unaware of it doesn't make it any less true.

Math is in your brain. It's a routine component of your thinking about anything and everything. BT just compels you to be honest about it and take seriously what that means. But mathematics itself is a difficult and foreign language. Most people are simultaneously bored and terrified by it.[18] So I will use numbers and formulas sparingly and simplify everything as far as I possibly can. In historical reasoning, this works well enough because we never have and thus don't need the advanced precision scientists can achieve (and indeed, many applications of BT can become exceedingly complex, especially in the sciences).[19] But that doesn't change the fact that the logic you always use when evaluating claims is inherently mathematical. Historians, by dealing with claims that very often can only be known to varying degrees of uncertainty, rely even more routinely on mathematical reasoning than the rest of us. That's why it's especially perverse for them to refuse to admit this or examine the actual logic of it.

When we introspectively examine how we intuitively estimate probabilities, we discover that we rarely think in absolutes. We all arrange our knowledge in cascades of different levels of confidence; some things we believe are more probable than others, and almost nothing we can say is absolutely 100% certain. Thus we already quantify our beliefs, even when we don't think exactly how or describe what we're doing using actual numbers. And again, this is especially true in history, where the data is often scarce and problematic, and thus our beliefs are often far less secure than when confronting the results of science or journalism or direct personal experience. So we shouldn't hide from these facts. BT simply describes, or 'models,' ideal reasoning about empirical probabilities. Like any logical syllogism, if you believe the premises, you must necessarily believe the conclusion, because the conclusion follows from those premises with deductive certainty. And those premises are the relative priors and the relative consequents: how much more likely (how much more typical) is one hypothesis than another on prior information, and how much more likely (how much more expected) is the evidence on one hypothesis than on another. And whether we're correct or not, or aware of it or not, we always have beliefs about what those relative probabilities are (as I'll prove in chapter 4; if you want to look at that now, jump to page 110).

This means if you don't follow BT, even intuitively, then you are not behaving rationally. You will be entertaining contradictory beliefs. And since BT describes the best way to reason, you will always reason better, and thus your beliefs will be more secure, when you follow BT, than when you just do the same thing intuitively, not really sure why your hunches are as they are or why your convictions should really follow from them. BT helps with this by exposing the numerical assumptions you are already making, and revealing their correct logical relations. And like any logical argument, since the conclusion (the final probability determined by BT) necessarily follows if the premises are true (the two priors and the two consequents), if your beliefs about what those premises are, are well-established or defensible, so is the conclusion.

"But math is hard." Another objection is that when using BT it's easy to screw up. There are many ways to err in Bayesian analysis—including numerous common fallacies in reasoning about probability.[20] Thus, to use BT competently it's important to get familiar with probability theory and the mathematics of probability and how not to err in applying it. But you won't avoid all those errors by avoiding BT. You will continue to make many of the exact same errors, only without being aware of it, whereas working with BT will force you to confront the possibility of these errors and so compel you to learn how to avoid them. Some errors, though, will be unique to using the language of mathematics. For example, all hypotheses you compare using BT must be incompatible (so that $P(h|{\sim}h) = 0$), and you have to attend to the correct means of differently treating independent and dependent probabilities. And so on. This is all statistics 101 and will be learned from any introductory college course or text. More advanced statistical techniques won't normally be of use to historians, so you needn't worry about them. But most errors are already commonplace even among those who have never heard of BT, such as developing overconfident priors or misestimating the likelihood of an alternative theory generating the same evidence. BT will actually help you catch these errors by exposing all the consequences of making them, and by forcing you to validly ascertain those probabilities instead—instead of pretending your reasoning isn't already relying on them (and often uncritically). Hence, avoiding BT will actually make your reasoning *more* susceptible to error. And in the end, "it's too hard" is not an argument we should ever hear from a professional historian—because mastering difficult methods is what separates professionals from amateurs. The bottom line is, if you're a historian, learning probability theory is your job.

"But history isn't that precise." A third worry is that math implies precision, yet in historical argument there can't be anything so precise, so using mathematical methods will give the false impression of precision where there is none. After all, you might feel justified saying something is "very probable," but that doesn't mean you know its probability is, say, exactly 83.7%. But the mistake being made here is assuming the one entails the other. You don't have to claim to know the exact probabilities in order to use BT. I'll discuss the mechanics of how to use inexact math in the next section.[21] For the purpose of BT is not to coax you into asserting precision you can't justify, but to correctly represent the logical consequences of the ranges of probability you *can* justify. Any uncertainty can be represented mathematically. And that uncertainty will validly carry over from the premises to the conclusion. In fact, that's the very merit of

BT: it correctly carries over all the uncertainties of your premises into your conclusion. BT can still be abused and misused or used incompetently or incorrectly. But identifying examples or possibilities of such abuse is not an argument against using BT, but against using it incorrectly.

It also isn't necessary that all historical writing and argument be mathematically formalized. It's only necessary that however historical claims are written and argued they be *capable* of transformation into a mathematical formalism—because that formalism represents the actual logic of any informally stated argument. The first section of this chapter, on the historicity of the sun going out for three hours, shows how a Bayesian argument can be articulated in plain English without any equations, numbers, or math. But to be checked and confirmed, it must be capable of being modeled by Bayes's Theorem, as was accomplished in a later section of this chapter. And when it is thus modeled, the conclusion must prove the same. Otherwise, the "plain English" will only have disguised a logically invalid or unsound argument. BT is thus a means of checking our work. We won't always have to show our work. But we should always be capable of doing that work, and doing it competently and correctly. And sometimes we *will* need to show our work, precisely so it can be checked and debated.

MECHANICS OF BAYES'S THEOREM

The following shall be the most math-challenging section of the book. It is essential to understand the math, because the math represents a logic, and this logic models the structure of all sound historical reasoning in every field of human knowledge (which I'll prove in the next chapter). The complete BT equation is again:

$$P(h|e.b) = \frac{P(h|b) \times P(e|h.b)}{[\,P(h|b) \times P(e|h.b)\,] + [\,P(\sim h|b) \times P(e|\sim h.b)\,]}$$

Here 'P' stands for 'epistemic probability,' and the symbol 'I' represents conditional probability, for example, $P(x|y)$ means the probability of x given y (i.e., what the probability of x is if we assume y is true). For example, the probability that a given person is named John *given that that person is a girl* is far lower than the probability that just anyone is named John. The former is a conditional probability. The variable h stands for 'hypothesis' (an explanation of the evidence we intend to test); e for 'evidence' (the evidence we intend to explain with h); b for 'background knowledge' (everything else we know); and $\sim h$ stands for all other hypotheses alternative to our own (all other possible explanations of the same evidence).

$P(h|e.b)$ thus means "the probability that our hypothesis is true, given the evidence and all our background knowledge" (in other words, "the probability that our hypothesis is true *given everything we currently know*"), and this probability follows necessarily from four others: $P(h|b)$, which means "the probability that our hypothesis would be true given only our background knowledge"; $P(\sim h|b)$, which means "the probability that our hypothesis would be *false* given only our background knowledge"; $P(e|h.b)$, which means "the probability that we would have all the evidence we actually do have, given all our background knowledge, *if* our hypothesis were indeed true"; and, finally, $P(e|\sim h.b)$, which means "the probability that we would have all the evidence we actually have, given all our background knowledge, *if* our hypothesis were instead *false*." From those four probabilities, the conclusion necessarily follows, which is the **posterior probability,** which in turn is simply the **epistemic probability** that our hypothesis is true.

If that "epistemic probability" is greater than 0.50 (i.e., 50%), then we have sufficient reason to believe our hypothesis (h) is more likely true than not, although our certainty will be attenuated to the actual value of that probability. So an epistemic probability of 0.90 leaves us far more certain that h is true than an epistemic probability of only 0.60, which would leave us very uncertain, leaning only slightly in favor of h being true, harboring considerable doubt. To calculate the epistemic probability for any h, we need to estimate only three values (from which the fourth automatically follows).

Each of these values is the equivalent of a 'premise' in a logical argument. Just as in any other logical syllogism, the conclusion (in this case, the epistemic probability we end up with) is never more certain than the weakest premise. Therefore, to apply BT correctly we often must allow for considerable degrees of error and uncertainty when assigning values to these three variables. Those variables are the **prior probability** your hypothesis is true (which is P(h|b)), the **consequent probability** of the evidence on your hypothesis (which is P(e|h.b)), and the **consequent probability** of the evidence on any other hypothesis (which is P(e|~h.b)). And from the first of these follows a fourth: the **prior probability** of any other hypothesis being true (or P(~h|b), which always equals 1 − P(h|b)). You can substitute for the last two of these premises the single premise P(e|b), as shown in the appendix (page 283), but that becomes less intuitive and more difficult for nonmathematicians to use correctly. You can also do the math in the form of "odds" instead of "probabilities" (page 284), but that often requires converting one to the other, an unnecessary step.

Sometimes you will want to take into account numerous hypotheses distinctly. For example, there may be two hypotheses competing against your own, one of which has a high prior probability but a low consequent probability, while the other has a high consequent but a low prior. Treating them together as a single competing hypothesis (~h) would thus be difficult to represent accurately, requiring you to tease out both and treat them separately. If you want to distinguish several competing hypotheses like this, you simply expand the equation, like this:

$$P(h_1|e.b) = \frac{P(h_1|b) \times P(e|h_1.b)}{[\,P(h_1|b) \times P(e|h_1.b)\,] + [\,P(h_2|b) \times P(e|h_2.b)\,] + [\,P(h_3|b) \times P(e|h_3.b)\,]}$$

If you want to test more than two alternatives to your own, just add as many boxes to the denominator as you need (e.g., "... + [P(h_4|b) × P(e|h_4.b)] + [P(h_5|b) × P(e|h_5.b)] ..." etc.). Just remember that all the prior probabilities in any expanded equation must still sum to 1. For example, if testing five hypotheses altogether (yours against four others), then you must ensure that $P(h_1|b) + P(h_2|b) + P(h_3|b) + P(h_4|b) + P(h_5|b) = 1$.

The mechanics of prior probability

The fact that all priors must sum to one is a useful aid to estimating priors. First you exclude all hypotheses with vanishingly small priors. For example, "space aliens did it" is always so inherently improbable its prior will surely be far, far less than even one in one hundred, indeed more on the order of one in a billion (or 0.000000001) or even less. So if there is no compelling evidence for that hypothesis at all, its effect on the equation will be essentially invisible. You can ignore it. Even the *sum* of the priors for every conceivable harebrained hypothesis will be substantially less than that. So unless there is specific evidence on hand for any of them, we can ignore them all.[22] That will result in your equation only producing an *approximation* of the probability that a given hypothesis is true, but that is all you need in historical analysis. We don't need precision to the tenth decimal place when the odds of an unlikely hypothesis being true are going to be a thousand or million times less than any more plausible hypothesis. So historians can simplify their labor by treating absurdly low probabilities as 0 percent and absurdly high probabilities as 100 percent (until they have reason not to—I'll say more about all this in chapter 6, page 249). That leaves you to deal only with the hypotheses that have more credibility.

If there is only one viable hypothesis, all others being crazy alternatives, then the sum of all the latter can become the prior probability of ~h as a catch-all alternative, and a very low probability it will be. But usually there are at least two or three viable hypotheses (or even more) vying for confirmation. Then it's only a matter of deciding what their relative likelihoods are, based on past comparable cases. How often are stories of miraculously darkened suns made up, relative to how often suns actually get blotted out? Even if you don't have other stories of the sun going out, you have comparable cases, such as tales of the moon splitting in two, armies marching in the sky, and crucifixes and Buddhas towering over the clouds. Adding it all up, you get a reference class (a procedure I'll discuss more in chapter 6), in which we find most of the comparable cases are 'made up' (or hallucinated or whatever else) rather than 'actually happened' (unless we agree that most of those cases are real, but then we must face the consequences of our now believing that giant space Buddhas visit earth and mysterious cloud armies might descend upon us at any moment). "Most" is a numerical assertion, especially in this context, where you certainly don't mean six out of ten such events are real and the other four made up. You will probably be quite confident that no more than one in one hundred or even one in a million of them could have been real. If you settle on the former, you have a prior probability that any such story is real equal to 0.01 and therefore a prior probability that any such story was made up (or merely records an illusion, delusion, or hallucination) equal to 0.99 (because these two options exhaust all possibilities, so we know the odds that one of these possibilities is true is 100%, and 100% − 1% = 99%).

I'll explain later why you might settle on that specific number. For now the point to be made is that priors must be assessed by comparing *all* the viable hypotheses against each other and deciding how likely each is *relative to all the others*—not in isolation from them. The biggest mistake amateurs make in determining priors in BT is to mistake the probability of an event happening with the prior probability of a story about that event being true. The physical probability that a giant Buddha will materialize in the sky is certainly astronomically low. But that's not the same thing as the *epistemic* probability that, when someone claims to have *seen* a giant Buddha materialize in the sky, they are neither lying nor in error. The priors in BT represent the latter probability, not the former. For only then will the prior probability of 'actually happened' and the prior probability of 'made up' (or whatever else) add up to exactly 100%, as they must do for any argument to remain logically valid.

The mechanics of consequent probability

Once you have your priors, you have to estimate the consequents. $P(e|h.b)$ represents how likely it is that all the specific evidence we have (everything included in *e*) would exist if our hypothesis (*h*) is true. In historical reasoning, this means the specific evidence that *h* (and its competitors) is meant to explain or has to explain. So if there is anything in *e* that we would *not* expect on *h*, then the consequent probability will be less than

1, in exact proportion to *how* unexpected the contents of *e* happen to be. Conversely, P(e|~h.b) represents how likely it is that all the specific evidence we have would exist if our hypothesis is *false*. But if *h* is false, then necessarily something else caused the evidence in *e*, and therefore some *other* hypothesis must be true. So we have to ask ourselves what the most likely alternative explanation actually is (or explanations, if several are plausible). Only then can we estimate how likely it is that the alternative(s) would generate the evidence we have.

These two probabilities don't have to sum to one. They can even be the same probability. They can even both be one. For if two hypotheses, h_1 and h_2, perfectly explain all the evidence—if all that same evidence would always exist on either hypothesis—then the consequents for both are indeed one (i.e., 100 percent). In such a case, there happens to be no evidence available that can tell the difference between them. So all we have to go on is what was typical in past cases, in which event the priors alone will tell us what's most likely. If, for example, a bunch of Tibetan peasants report seeing a giant Buddha in the sky and there is no way to test that claim against any other evidence, we have to conclude they either hallucinated *en masse* or fabricated the story (or are delusional or victims of an optical illusion, etc.), not because the evidence of the case verifies this conclusion (we would have exactly that same evidence—their mere report—whether the story was true or false), but rather because that's the most inherently probable explanation. That's why extraordinary claims require extraordinary evidence: to overcome the overwhelming prior probability against such claims being true.[23]

Estimating consequents is simply a question of asking yourself some questions about what each plausible hypothesis actually predicts. Begin by asking yourself if the evidence would be any different if your hypothesis were false. Then ask *how* different it would be and assign that a value in terms of likelihood. If the evidence wouldn't be any different at all, then the *alternative* consequent, (P(e|~h.b)), equals one. If, on the other hand, the evidence would certainly be very different, then the alternative consequent must necessarily be far *less* than one. For example, consider our hypothesis that the darkened sun story was made up (or merely records the hallucination of a few fanatics): if that hypothesis were false, then the sun really did go out, in which case the evidence would be vastly different indeed (we would have nearly worldwide attestation in countless sources). The alternative consequent would therefore have to be very low (I assigned it a value of one in one hundred, and that was being absurdly generous).

Next, ask yourself if the evidence could actually be *better*. Could your hypothesis be even *more* confirmed than it already is? Such evidence, if you had it, would lower the alternative consequent *even more*. Accordingly, the *absence* of such evidence must be reflected in allowing the alternative hypothesis a *higher* consequent probability than you might otherwise have assigned. It may seem counterintuitive, but the best way to increase the probability of your theory being true is to *decrease* the probability of the evidence on every other viable theory. That's the actual effect of finding and presenting more evidence: such evidence makes alternative explanations less likely, hence making your explanation more likely. If even a single item of evidence is much less likely on any theory but yours, then that's what it means to call that item "very good evidence." The harder it is for alternative theories to explain that evidence, the stronger the support it gives to your theory. Likewise, probabilities accumulate; hence, the more evidence you have, if each item individually is less likely on alternative theories, then having *all* of that evidence is much *less* likely. It may be unlikely to have one eyewitness attesting the darkening of the sun, for example, but to have ten independent eyewitnesses doing so would be ten times less likely—not in a strictly literal sense (the actual math would vary from case to case) but it would involve ten acts of multiplication, each of which reducing the probability further. Thus both the quality *and* quantity of evidence are accounted for in BT.

Then turn the tables. Ask yourself whether there is any evidence that is what an alternative hypothesis would predict, but that your hypothesis doesn't, or at least not as well. If there is, then *your* consequent (P(e|h.b)) must be reduced to reflect how unlikely that coincidence would be if your hypothesis were still true. If there is a forged letter among your evidence, and your theory doesn't explain why it's there, but a competing theory does, and if both theories explain all the remaining evidence equally well, then your consequent must be less than the other consequent, which means your consequent cannot possibly be one, so it must be lowered to reflect the fact that the evidence you actually have is at least somewhat unexpected. And the other consequent must then be higher, as much higher as reflects how much more likely the evidence is on that alternative explanation.

This can get complicated, because reality is often not so black and white. For example, if you hypothesize that your neighbor is trustworthy but you discover he has a criminal record, that is not something your hypothesis would predict, but it is something the alternative hypothesis (that your neighbor *isn't* trustworthy) would predict (predict, that is, as having at least a higher than average probability, which is why, if you found him untrustworthy and then discovered he had a criminal record, you wouldn't be surprised). And yet, that alternative theory does not *entail* your neighbor will have a criminal record (he can still be untrustworthy without having a criminal record). His having such a record is just *more likely* on that theory—whereas it is less likely on the theory he is trustworthy. This is because, of the two classes of people, the trustworthy and the untrustworthy, fewer in the former class have criminal records than members of the latter class do. If no trustworthy people had criminal records, your hypothesis would be all but refuted by the discovery of a criminal record— your consequent would be as low as represents the merest remaining possibility that there may yet be some exceptional person not yet documented who has a criminal record and is still trustworthy. So if that were the case, your theory's consequent would be greatly reduced indeed. In reality, though, many people with criminal records are nevertheless trustworthy, so it would not be reduced quite so far, only as far as represents the actual likelihood of such a person still being trustworthy. On the other hand, the *lack* of a criminal record is not unexpected on the alternative hypothesis that your neighbor is untrustworthy, since many untrustworthy people lack criminal records. Thus the logic of evidence is often not as straightforward as many think.[24]

Assume (merely for the sake of argument) that the following represents the statistically determined facts (by, say, a very large scientific study):

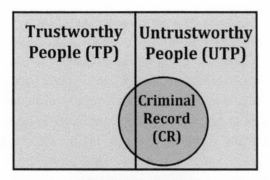

P(CR|TP.b) = 0.01
P(CR|UTP.b) = 0.08

P(~CR|TP.b) = 0.99
P(~CR|UTP.b) = 0.92

By this account, the consequents of *both* hypotheses would be small. But it is only the ratios that matter to the outcome. And in this account, having a criminal record is *eight times more likely* on the "untrustworthy" hypothesis than on the "trustworthy" hypothesis, whereas *not* having a criminal record makes very little difference on either hypothesis. Hence, the absence of a criminal record reduces the consequent for ~*h* (untrustworthy) by only a tiny amount (and for *h*, an even tinier amount), whereas the presence of a criminal record reduces the consequent of *h* (trustworthy) eight times more than the consequent of ~*h*, the same as if the consequent for ~*h* were 1 (100 percent) and the consequent of *h* were 0.125 (merely 12.5 percent), which is a huge difference.

In such a way, evidence can reduce the consequent probability of your hypothesis, and sometimes reduce it greatly, even though that evidence doesn't even contradict your theory—just as having a criminal record and being trustworthy are not mutually contradictory. As long as you lower your consequent to reflect the fact that some of the evidence is *less* expected on your theory than alternatives, your reasoning will be sound. But if you don't take this into account (and historians who avoid BT often do not), your reasoning will be fatally flawed. Thus using BT will often uncover errors otherwise overlooked, such as ignoring the effect of different *degrees* of fitness between evidence and theory—rather than considering only evidence that directly 'contradicts' your theory as counting against it (a mistake too many historians make). Remember, *you may already be making this mistake*. So you can't avoid it by avoiding BT.

In fact another way to analyze consequent probabilities is simply to think of them as in ratio to each other (see page 284). As in the above case, all that really mattered was that P(CR|UNTRUSTWORTHY) = 8 × P(CR|TRUSTWORTHY), since no matter what probabilities you use for P(CR|UNTRUSTWORTHY) and P(CR|TRUSTWORTHY), as long as this ratio of eight to one is maintained, the result using BT will always be the same. Thus you don't need to know these probabilities at all, just what the ratio between them is. Sometimes knowing the latter even tells you the former. For example, if several witnesses report seeing a series of bright lights in the sky moving relatively quickly and changing color but staying in formation (and they saw nothing else and no other evidence turns up), there are three likely explanations: aircraft flying in formation, an aerial flare drop, or a meteor breaking up in the atmosphere. If only one in a hundred meteors breaks up (and thus looks like what these witnesses reported), and aircraft never look like that (never being that bright), but flares almost *always* look like that, then you can say the evidence is a hundred times more likely on the "flares" hypothesis than on the "meteors" hypothesis, and millions of times more likely on the "flares" hypothesis than on the "aircraft" hypothesis. If flares always look like that, then P(e|FLARES) = 1; and if aircraft produce that same evidence millions of times less often, then P(e|AIRCRAFT) < 0.000001 (in other words, one in a million at best); and if meteors produce that evidence only a hundred times less often, then we have P(e|METEOR) = 0.01 (or one in a hundred). But whether thinking in ratios like this, or analyzing the problem in any other way, the question is always the same: how often does a particular cause (*h*) produce the kind of evidence you have (*e*)? That's the consequent probability.

Consequent probability and historical contingency

Like the future course of the Mississippi River, predictions entailed by a hypothesis are always to some extent (and in some cases entirely) generic rather than specific. This is especially the case in history. The resulting probabilities are always conditional on a large set of other hypotheses. Most of these consist of 'background knowledge' (regarding the nature of the world, and of people, cultures, and contexts), which are hypotheses so well confirmed that their probability of being true is well above 99 percent. But some are hypotheses regarding historical contingencies beyond our ken. For example, in science a hypothesis may predict what observation will be made in what contexts, but it doesn't predict exactly when and by whom. Thus we can safely ignore such specific details as "exactly when and by whom" when estimating consequent probabilities. The same follows for hypotheses about history.

Specifying the 'type' of evidence to expect in this way allows wide ranges of possible outcomes, such that any one of those outcomes can be accounted likely if it occurs. Thus the probability that an observation would be made by a specific scientist on a specific date, which probability will always be extremely low, can be left out of account. So if A, B, C, etc. are different scientists making the same kind of observation and on different dates, a hypothesis might predict that P(A or B or C . . . *ad infinitum*|h.b) = 100%. In other words, even though P(A|h.b) = 0.000 . . . 0001 (since the probability of that observation being made by that specific scientist on that specific date will be exceedingly small), we are only interested in P(A or B or C . . . *ad infinitum*|h.b). Thus *h* does not predict *exactly* what evidence will appear, only what *type* of evidence will appear (what *sort* of thing *some* scientist will see on *whatever* date).

History entails many other kinds of disregarded contingencies like that. For example, that the Gospel according to Mark came to be attributed to "Mark" rather than to, let's say, "Timothy," may be a mere accident of history that makes no difference, since a particular hypothesis might only predict that there would be some such document *whatever the attribution*. Likewise the specific content of such a document: there are countless different ways the Gospel of Mark could have been written, countless different stories and constructions and word choices and sentence structures. Most hypotheses don't predict these exact details, only what *sort* of contents should be there. The same goes for all evidence of any kind. One must thus distinguish 'predictions of exact details' (which BT does not concern itself with in this case) from 'predictions regarding the type of evidence to expect.' I'll provide a more technical demonstration of the validity of this distinction in chapter 6.

But you must be certain the issue is of the expected type and not a required specificity. In some cases, a disregard of contingency, such as expressed by $P(A \text{ or } B|h.b) = 100\%$, will not be valid. As with the example of the trustworthy neighbor: defining e in such a way as to entail $P(CR \text{ or } \sim CR|h.b)$ produces invalid results, since both consequents would then be 100 percent (because the probability that anyone, trustworthy or not, would either have or not have a criminal record is by definition 100 percent), when in fact the presence of CR (a criminal record) makes a huge difference in the consequents (and the absence of CR, a small one). So when a particular contingency becomes more likely on one hypothesis than another, it can no longer be disregarded. For example, one hypothesis might predict that the Gospels would all be attributed to a disciple (and Mark, not being a disciple, would thus be an unexpected attribution), in which case $P(e|h.b)$ would have to be lowered due to the evidence not being as expected. Or it could even be the other way around. When near enough to the events related, inventing a nonexistent author is always safer and thus more likely on a hypothesis of fabrication, because a nonexistent author won't be around or have living relatives who could gainsay the claim that he'd written any book, much less that one—hence, on a hypothesis of fabrication within that time frame, we should more expect attributions to people who can't be precisely identified or who don't even exist, in which case $P(e|h.b)$ would be raised, not lowered. But either way we have a differential prediction we must account for. Of course, that difference in probability might also be small. It might even be so small as to be washed out by an *a fortiori* assignment of probability (which I'll discuss later) and therefore ignored on that ground instead. Likewise, such a difference in consequents might only occur for hypotheses whose prior probability is vanishingly small, which are already thus disregarded on *that* ground (as I explained on page 70). Otherwise, when two hypotheses make different predictions regarding the exact contents of *e*, as long as both hypotheses are viable (and thus not excluded on priors) and the difference predicted is large (and thus has a significant effect on consequents), the distinction must be included in your estimates of probability.

The role of conditional probability

All these probabilities are conditional on *b* (our background knowledge). Thus, just as with our priors, we base our estimates for the consequents on what we know about the world, people, the culture and historical context in question, and everything else. This means there are four ways to misuse the evidence. You can put things in *b* or *e* that shouldn't be there, or fail to include things in *b* or *e* that should be there. Both *b* and *e* should only contain facts every reasonable expert can agree on. Contentious claims, speculative suppositions, hypotheses that can't produce an expert consensus, and things that aren't even true should not be considered as either evidence or background knowledge when estimating probabilities in BT, although the mere fact that an expert disagreement persists on some point can *itself* be reckoned as knowledge or evidence. Background knowledge ideally represents the established consensus of experts, which can include a consensus that there is no consensus; while the evidence represents the facts that everyone agrees need explanation.

This also means you cannot *exclude* facts of either kind. Abusers of BT often attempt to argue in a vacuum, pretending a great many things we all know (the complete contents of *b*) aren't known, merely to generate a result they like. If we *know* a vast number of miracle claims have been established to be fraudulent, erroneous, or inaccurate (and we do), we cannot pretend otherwise. We *must* take this into account when estimating the prior probability of a genuine miracle. Because even if we are personally certain that some miracles are genuine, we still know for a fact most miracle claims are not. Therefore, we must accept that the prior probability of a miracle claim being true must still be low. Likewise, BT can be abused by excluding facts from *e* that substantially affect the consequent probabilities. The fact that the early growth of Christianity was exactly comparable (in rate and process) to other religious movements throughout history is a fact in evidence that significantly challenges the claim that Christianity had uniquely convincing evidence of its promises and claims.[25] Likewise, the fact that medieval Christians became as depraved and despotic as peoples of any other faith is unlikely on the hypothesis that they had any more divine guidance or wisdom than anyone else has ever had. Similarly, the frequency of admirable new ideas among them was no greater than among many other cultures (who came up with many admirable ideas of their own), so they cannot claim any greater inspiration, either.[26]

Thus any fact in evidence that is hard for your theory to explain, or that is more likely on another viable theory, must be included in *e*. This also means it is not logically valid to exclude evidence from *e* by the mere device of inventing an excuse for it. Because any such excuse must necessarily lower the prior probability of your original theory, due to the simple fact that a theory without that excuse must have a nonzero prior—because it is logically possible, and indeed may even be the more plausible—so a theory *with* that excuse must have a prior that is *smaller*. This is because the prior probability of either theory being true must equal the sum of the prior probabilities of each one being true separately. For example, if theory A includes explanation D *and* excuse C, and theory B also includes explanation D but excludes excuse C, the prior probability of 'explanation D' (with or without C) equals the sum of the prior probabilities of A and B. In other words: $P(D) = P(A) + P(B)$.[27] So if $P(B)$ is any nonzero value (and we know it always must be), then $P(A)$ must be less than $P(D)$ by exactly that amount. In other words: $P(A) = P(D) - P(B)$. Thus adding C always reduces the prior probability of D. In fact, unless you have evidence supporting the inclusion of the excuse (C) over its exclusion (which means evidence *other* than the evidence that you invented C to explain away), then including C necessarily *halves* the prior of D—since with no evidence confirming C or ~C the *principle of indifference* entails that the probability of either must be equal (because for all you know either is as likely as the other), which means a probability of 0.5. All probabilities must be conditional on what you know at the time, so if you know nothing as to whether it would be higher or lower, then so far as you honestly know, it's 0.5. And if there is evidence *against* C (either against C specifically or against excuses like C generally), then including C will reduce the prior of D by *more* than half. Since then the probability of either C or ~C is no longer equal but tilted in favor of ~C. This iterates for every excuse added. Thus, attempting to salvage a hypothesis by inventing numerous *ad hoc* excuses for all the evidence it doesn't fit will rapidly diminish the probability of that hypothesis being true (which happens to be the logical basis for Ockham's Razor—as I'll explain in the next chapter, page 104). Only if you can adduce convincing evidence that such an excuse was indeed operating will including it have a negligible effect. For instance, if you can prove that D *without* C is actually very rare or unlikely, then P(A|b) will be close enough to P(D|b) as to make no significant difference to the outcome. And either way, we still don't remove the evidence thus "explained away" from *e*; we merely assign that evidence a high consequent on *h* (since C will have been crafted to have exactly that effect: to make the evidence more likely on *h* than would be the case without C). This is what prevents 'gerrymandering' a cherished theory to fit just any body of evidence. Which makes BT an important corrective to bad historical reasoning.

The problem of subjective priors

The objection most frequently voiced against BT is the fact that it depends on subjectively assigned prior probabilities and therefore fails to represent objective reasoning. The same can be said of the consequent probabilities. But insofar as this objection is true, it condemns *all* reasoning, not just BT. The only difference is that BT makes this reliance on subjectively assigned priors apparent, whereas all other methods of argument simply conceal it. The fact remains that we always base our arguments on assumptions about the inherent likelihood of whatever we are arguing for or against. If that is a defect that condemns any method, it condemns them all.[28]

But the subjectivity of priors is actually not a problem for BT.[29] The fact of their subjectivity does not prevent us from producing conclusions that are as objective as can possibly be, given the limits of our knowledge and resources. Because subjective does not mean arbitrary.[30] You must have *reasons* for your subjective estimates, so you must confront what those reasons are.[31] And in defending your conclusion to anyone, you must be able to present those reasons to them, and those reasons had better be widely convincing. If they aren't, you need to ask why they ever convinced you. If you have no such reasons, then your priors are irrational and you need to change them until they are rationally founded. And if you do have sufficient reasons, you need to ask if those reasons will be accepted by other reasonable people. That would then actually make them objective, since by definition objective reasons will be accessible and verifiable to all reasonable observers, who will thus all come to the same subjective estimate.

You can still warrant your own belief in a conclusion if you have access to evidence others don't, but in such a case you need to acknowledge and accept the consequences of that fact. One of those consequences is that whether others accept your testimony will be based only on the evidence available to *them*. For example, you may be certain you visited an alien spaceship last night, but everyone else only has a vast body of background knowledge establishing that most (if not all) such experiences are hallucinatory or fabricated. You have to respect the fact that they are fully warranted in *rejecting* your testimony—until such time as you can prove your experience was genuine in a way that so many others were not, with evidence that others *can* observe (I'll discuss this point further in chapter 6, page 210). Likewise, if you accept or reject different epistemological assumptions, then you cannot persuade anyone with *any* argument (much less BT) until you either adopt their epistemological assumptions or a subset of them that you share in common (enough to build a valid argument on), or first convince them to adopt your epistemology instead. None of those scenarios are relevant to the present case. This book is not written to convince adherents of radical epistemologies (such as reject the axioms and rules discussed in chapter 2). And like all professional historians, I am only interested in objectively grounded premises (in accord with the first axiom).

So how do we get those? You may be hesitant to assign probabilities because it seems arbitrary and subjective. But the point is to translate your actual beliefs into a more convenient language. You already have those beliefs. So translating them into numbers does not make them any more arbitrary and subjective than they already are. And they are rarely as subjective as you might think. When you say a claim is plausible, you are saying it has a high enough prior probability to consider it. When you say it's implausible, you are saying it has a prior probability low enough to dismiss it. Whenever you say one theory explains the evidence better than another, you are saying it has a higher consequent probability than another. And so on. All these statements will entail numerical equivalents, which are often objectively reasonable.

Given any belief about the past, you must believe it has *some* probability of being true or false. And given any probability P, you logically *must* believe P is either 0 or 1 or some value in between. Only if the claim assigned this probability cannot be denied (if its truth is logically necessary) can P = 1, and only if it literally cannot be true (if its truth is logically impossible) can P = 0, because everything else has some nonzero chance of being true or false, as explained in chapter 2—and as also explained there, even those rare assignments of 1 and 0 cannot really be warranted, since we can sometimes be wrong about what's logically necessary or logically impossible. So nearly *every* claim has some probability between 0 and 1. Therefore, if P pertains to any claim about history, then P *must* be some value between 0 and 1. Where between? If you genuinely have no reason to believe P is higher or lower, then that *entails* that you believe P is 0.5. The latter is simply a translation of the former into a different language. If you disagree with that conclusion, then you either do so irrationally or rationally. If irrationally, then you are no longer participating in valid historical argument. You can safely put this book down. We have no use for you. But if you have a rational objection to the conclusion that P is 0.5, then you *must* have a valid reason to make P higher or lower, in which case you should raise or lower P accordingly. This is true by definition. If you have any objective reason to believe P is not 0.5, then you *must* believe it is either higher than 0.5 or lower than 0.5.

And when your language gets more precise (and you start using adverbs like "slightly," "very," or "extremely"), this, too, entails numerical equivalents. For example, "slightly more likely than 50%" never means 90%, or even 70%; typically it means no more than 60% and in some contexts might mean no more than 51%. But even if for some strange reason you actually mean by "slightly more likely than 50%" a 90% likelihood, then that is what it means to you. So you still have a probability. For any ordinal ranking of likelihood, there is some probability that you already mean by it, into which it can be translated. This won't be the precise and only probability you can mean by it, but there will be an upper and lower bound of the range of probabilities that you mean by it, and you can use either depending on which you need for the occasion (whether the lower bound or the upper) in order to build an argument *a fortiori* (which method I shall discuss on page 85). And the same goes for any other ordinal assignment, like "very likely" or "almost certain" or "quite probably" and so on down the line.

BT or not, either you could back up all these adverbs and assertions when you used them before, or you couldn't. If you couldn't, then you were being as subjective as you could ever accuse BT of being. But if you could back them up, then you can in BT just as well. So BT is no more subjective than any other form of historical argument. And when used properly, it is as objective as any historical method ever could be. Historians all have some idea of what was typical or atypical in the period and culture they study, and they can often make a case for either from the available evidence. If they can't, then they must admit they *don't know* what was typical, in which case they can't say one theory is initially more likely than another. The prior probabilities are then equal. Likewise, if historians can't defend a particular estimate of consequent probability, then they need to lower or raise that consequent until they *can* defend it. And if they can't defend *any* value, then they cannot claim to know whether the evidence supports or weakens their hypothesis (or any other hypothesis for that matter). *This follows whether historians use BT or not.* Because if they can't defend any probability assignments in BT, then its probabilities are all 0.5, and then its conclusion is necessarily always 0.5, which entails the theory in question only has a 50/50 chance of being true (or, for that matter, false). Since BT is formally valid and its premises would then be inarguably sound (because being conditional on what the historian knows, that's exactly what those probabilities must be if the historian can adduce no evidence for any other values), any argument that contradicts that conclusion (that the theory in question is as likely false as true) *must* be invalid or unsound. Likewise for any values other than 0.5 that a historian can make any defense of. The consequences of this are laid out in the next chapter.

So you might as well use BT. Because you can't get any better result with any other method.

Arguing *a fortiori*

There are several tricks to ensure your use of BT is adequately objective. The most important is employing estimates of probability that are as far against your conclusion as you can reasonably believe them to be. This is called arguing *a fortiori*, which means "from the stronger," as in "if my conclusion follows from even these premises, then my conclusion follows *a fortiori*," because if you replaced those estimates with ones even more correct (estimates closer to what you think the values really are), your conclusion will be even more certain.

This eliminates the problems of **inscrutability** and **underdetermination**. The exact probabilities in BT will often be inscrutable (meaning incalculable or unknown to us), but *a fortiori* probabilities will not be inscrutable. Their assignment will often be objectively undeniable. You might not personally know what the exact probability is of a mile-wide asteroid striking the earth tomorrow, but you certainly know it is less than one in one hundred (in fact, a very great deal less, since not even one such asteroid hits the earth every year, much less three of them). Thus the inscrutability of the *actual* probability does not entail the inscrutability of an *a fortiori* probability. Any conclusion reached with BT using such a probability will commute the *a fortiori* qualifier from the premises to the conclusion; for example, if we assign a prior probability of "*x* or less," then the conclusion will be some value "*y* or less," in other words, "possibly less, even a great deal less, but certainly not more." Thus we can establish a conclusion objectively with BT even when the actual probabilities are inscrutable.

Likewise, underdetermination refers to the problem that there are infinitely many theories that can explain all the same evidence just as well as ours does, and most of them we haven't even thought of (so we can't even say we've ruled them out). But since BT conclusions are conditional on present knowledge, theories we haven't thought of yet make no difference to the result. Such theories do not exist in *b* and thus have no effect on what we are entitled to believe given what we so far know. When we know something different, our conclusion about what's warranted may change, but only then (see chapter 6, page 276). And of theories we *have* thought of, most by far have a prior probability we know to be vanishingly small (like the ubiquitous "aliens did it"), while the remainder we consider. That's why adding up the priors of all the thousands of conceivable fringe theories will still only get us a total much less than one percent. Strictly speaking, we would still have to plug that into our equation. However, if we are using *a fortiori* estimates of the prior probability of our favored theory, then the collective priors of all conceivable fringe theories are washed out, buried under the much more enormous percentage by which we have already underestimated our own theory's prior. That's why we don't have to give them any further thought (unless we want to).

Arguing *a fortiori* can also save us a great deal of labor or help us identify where more labor would be worthwhile. The vast effort involved in trying to collect and analyze all the data necessary to get increasingly accurate estimates of probability is unnecessary if we don't *need* increasingly accurate estimates. If we can fully justify an *a fortiori* probability on a sound representative sample of the data (or an adequate preliminary survey of it) such that we can demonstrate that any continued labor will only push the result even further in favor of our conclusion, and yet our conclusion is already more than adequate to warrant confident belief, then we don't need to continue the inquiry further (unless we want to).[32] Conversely, if an *a fortiori* conclusion is not yet strong enough to warrant confidence, it nevertheless shows us in what direction further data will take us, thus giving us a reason to pursue that further inquiry precisely to see how much more confidence we can warrant in the conclusion.

Arguing *a fortiori* in BT also answers the objection that historians don't have precise data sets of the kind available in the sciences. All probabilities derived from properly accumulated data sets have a confidence level and a margin of error, often expressed in various ways, like "20% +/-3% at 95% confidence," which means the data mathematically entail there is a 95% chance that the probability falls between 17% and 23% (and, conversely, a 5% chance that the probability is actually higher or lower than that). Widening the margin of error increases the confidence level according to a strict mathematical relationship. This permits subjective estimates to obtain objectively high levels of confidence. If you set the margin of error as far as you can reasonably believe it to be, then the confidence level will be as high as you reasonably require. In other words, "I am certain the probability is at least 10%" could entail such a wide margin of error (e.g., +/-10% on a base estimate of 20%) that your confidence level using that premise must be at least 95% (a confidence level that with most scientific data sets would entail a much narrower margin of error). Again, you may not have the data to determine an exact margin and confidence level, but if you stick with *a fortiori* estimates, then you are already working with such a wide margin of error that your confidence level must necessarily be correspondingly high (in fact, exactly as high as you need: i.e., if the margin is "as wide as I can reasonably believe" then it necessarily follows that the confidence level will be "as high as ensures my belief is reasonable"). Thus precise data is not needed—unless no definite conclusion can be reached with your *a fortiori* estimates, in which case you have only two options: get the data you need to be more precise, or accept that there isn't enough data to confidently know whether your theory is true. Both will be familiar outcomes to an experienced historian.

Indeed, "not knowing" is an especially common end result in the field of ancient history. BT indicates such agnosticism when it gives a result of exactly 0.5 or near enough as to make little difference in our confidence. BT also indicates agnosticism when a result using both margins of error spans the 50% mark. If you assign what you can defend to be the maximum *and* minimum probabilities for each variable in BT, you will get a conclusion that likewise spans a minimum and maximum. Such a result might be, for example, "45% to 60%," which would indicate that you don't know whether the probability is 45% (and hence, "probably false") or 60% (and hence, "probably true") or anywhere in between. Such a result would indicate agnosticism is warranted, with a very slight lean toward "true," not only because the probability of it being false is at best still small (at most only 55%, when we would feel more confident with more than 90%), but also because the amount of this result's margin falling in the "false" range is a third of that falling in the "true" range. Since reducing confidence level would narrow the error margin, a lower confidence would thus move the result entirely into the "true" range—but it would still be a very low probability (e.g., a 52% chance of being true hardly instills much confidence), at a very low confidence level (certainly lower than warrants our confidence).

This method of using *a fortiori* probabilities was illustrated earlier in this chapter with both examples of the sun going out. Exact data were not necessary to determine what conclusion is obviously correct. The frequency of such claims (of wildly unlikely astronomical events) being true (rather than, for any of various reasons, false) is certainly far less than one in one hundred, which entails the probability that the sun went out in the first century is certainly far less than the one in ten thousand we concluded with. That conclusion is objectively true. Yet at no point did we need to know the *exact* ratio of true to false claims in the category of wildly unlikely astronomical events.

Mediating disagreement

There is no rational basis for rejecting the validity of BT, and, as so far shown, no reasonable basis for rejecting its application to history (and chapter 4 will prove that), which leaves disagreements over its application. These will consist of objections to the correctness of its use (which ought to be resolved not by abandoning BT but by using it correctly) and objections to probability assignments and their derivation from accepted background knowledge. The latter are exactly the kinds of debates historians need to be engaging in. Because by avoiding BT they are often avoiding the *real* issues of contention between them, and thus failing to address the actual defects of their own methodologies. The result is a chaos of opinions battling each other as claims to fact, with no progress in sight (such as surveyed in chapter 1). But if the difficulty of subjectively arriving at objective probability estimates is confronted head-on, then progress can be made. Even if that progress only amounts to a greater consensus on our mutual uncertainty, it would still be progress.

Most disagreements arising from the subjectivity of probability estimates are of no relevance to progress in the field anyway. For example, if two scholars disagree over the correct prior, one insisting it must be a thousand times smaller than the other is sure it is, yet on either estimate the conclusion is still more than 99 percent certain (i.e., P(hle.b) > 0.99), then their disagreement makes no relevant difference. Both historians would consider the hypothesis decisively confirmed. Only when disagreements over probabilities make for a different conclusion do those disagreements matter. And when that happens, each scholar is obligated to present the evidence establishing that his estimate is correct and that the other is not. Often such an exchange will conclude with both scholars agreeing on an *a fortiori* estimate somewhere in the middle as the only estimate that can be objectively supported by the evidence. The result will be agreement on the estimated probabilities, and, therefore, agreement on the conclusion. This is one of the reasons historians should use BT: precisely because it will provoke such debates and discussions, resulting in greater clarity, better-supported premises, and wider agreement.

Indeed, to begin making progress, you have to start somewhere. Imagine a situation where all observers start with different estimates and then acquire exactly the same total knowledge; that is, each observer knows exactly all the same things as every other observer. There could then be no explanation for why they would still differ in their estimates, because, if they only employ valid arguments, working from exactly the same information, then they would have to agree. It would be logically impossible for anyone to reach a different conclusion under those conditions. So if they still came to a different conclusion, we would have to identify irrationality as the culprit, and then work out who is guilty of it by examining ("on the couch," so to speak) why each observer still comes to different estimates. I think in practice, as knowledge sets converge, rational observers (those who only employ logically valid arguments) will converge in their probability estimates, and even where they continue to differ, the difference will become less significant (as in the manner described above). One scholar might disagree with another yet still accept the other's estimates as approaching what's reasonable, especially if their margins of error overlap. Hence, they might consider their differences too insignificant to signal either of them as irrational (unless they can identify actual instances of invalid reasoning).

More likely any outstanding disagreements would be due to remaining differences in their knowledge sets, especially knowledge not easily communicated or shared (such as personal childhood memories and influences). Even when that difference comes from years of study in a particular area that one historian has undertaken but others have not, since that will still have been the study of actual objective facts, that historian should still be able to collect and communicate any relevant data culled from it whenever claims based on it are doubted. Thus, if significant differences in estimates exist between experts, the solution is not to abandon BT (since those differences will remain and covertly influence all their arguments anyway), but to increase communication through debate and information sharing so their respective bodies of evidence and background knowledge approach agreement and so their reasoning will approach greater and more consistent validity. Which, of course, historians should be doing anyway.

The conclusion must be that if you cannot validly formulate your argument according to Bayes's Theorem, then either you don't know what you're doing or your argument is invalid or unsound (as we'll see in the next chapter). Therefore, a good method is to attempt such a formulation for any argument you may ever make as a historian and then identify what problems arise with the attempt. Because those problems will often expose underlying assumptions you had already been making or relying on that are not as sound or secure as you may have assumed. Or, you will discover how you are misusing BT, and by correcting that mistake you will thus improve your competence in applying it.

I will say more on the matter of resolving disagreements with BT in chapter 6 (starting on page 208). But ultimately, any argument against the

applicability of BT in any given case can only amount to either: (1) a declaration that the data is too scarce to come to any conclusion (in which case
a proper use of BT will prove this and thus BT will not be inapplicable
after all); or (2) a declaration that BT has been misused (in which case
the proper response is not to abandon BT but to use it correctly, so this
cannot be an argument for its inapplicability); or (3) a declaration that the
user has employed the wrong data, either leaving something out of account
or including something false or inappropriate, in which case, again, the
proper response is to redeploy BT with the correct data (which will produce
a conclusion all rational parties must agree with, so even this cannot be
an argument for its inapplicability). In terms of (1), (2), or (3), continued
honest debate will lead to agreement by all rational parties as to the premises—regarding allowable margins of error, for example.[33] Once everyone
agrees upon the premises, they must agree with the conclusion, as it then
follows necessarily. If, despite all this, someone continues to insist upon the
inapplicability of BT, there is not likely to be any rational ground for his
opposition. Usually at this point objections to BT amount to a stolid refusal
to accept rational conclusions that one is emotionally or dogmatically set
against, which is definitely *not* a valid objection to BT's applicability.

One might still object not to BT's *applicability*, but to its *utility*. That
is, we can acknowledge that BT is valid and that BT arguments in history
are sound, while still claiming that BT adds no value to already-existing
methods, or even makes things harder on the historian than they need to be,
in terms of both learning curve and time-in-use. But this objection will be
more than amply met in the following chapters. In general, there are three
points to make. First, insofar as historians are doing their job as professionals, they should already be devoting considerable time to mastering the
relevant methodologies. And yet learning BT is no more time-consuming
than learning existing methods of historical reasoning, and as existing
methods are significantly flawed (as will be shown in coming chapters),
the time spent learning the latter can be redirected toward learning the
former (or rather, new versions of the latter that have been reformulated
in terms of the former), resulting in no net loss in learning time. Second,
insofar as historians are *not* spending time learning logical methods of

analysis and reasoning, they *ought* to be, otherwise they cannot claim to be professionals. Complaining that learning how to think properly is too time-consuming is not a complaint any honest professional can utter in good conscience. Third, when it comes to time-in-use, in most historical argumentation BT does not have to be very complicated or time-consuming at all, except in precisely the ways all sound and thoughtful research and argument must already be. And in the few cases where BT arguments need to be more complicated, this will be precisely because no other methods exist to manage those cases. Because if they did, and they are logically valid, then they will already be covertly Bayesian anyway (as I'll demonstrate for a number of cases in chapter four). When a problem is complicated, it will always be complicated no matter what tools you use to analyze it. Attempts to avoid that fact can only result in lazy or unsound thinking, and that is certainly not a good excuse to avoid BT.

Which brings me to a final point about the function of this book as a whole. I am pursuing two related objectives: first, to demonstrate when and why existing methods of historical reasoning are logically valid; and second, to provide a model of reasoning that can be directly employed in historical analysis and argument. The latter is methodological, the former is epistemological. But even the epistemological point is essential to vetting and developing new methodologies. Thus I shall explain in coming chapters how existing methods conform to Bayes's Theorem and only remain valid to the extent that they continue to conform to Bayes's Theorem. This then gives us warrant to trust those methods (by grounding them logically) even if we don't explicitly employ BT when we deploy them. That same analysis will also establish limits and guidelines for any effort to expand, improve, or modify existing methods, or develop new ones. But this still means that once a method's soundness has been established by BT, you don't have to use BT. You can simply apply the method it has validated. BT would remain a tool for criticizing when that method is being used invalidly, but as long as that method is used validly, its connection to BT need no longer be discussed.

So BT is not necessarily a *replacement* for other methods, but just the structure on which they must rest. And yet BT remains a very useful tool in its own right. As shown by the opening example of this chapter, most historical reasoning already implicitly conforms to it. By seeing this and rendering it explicit, casual reasoning becomes more refinable, testable, and critical. This often requires no math at all, but just an understanding of how the relative weights of certain probabilities entail a conclusion, and thus which probabilities you need to look at, and which weights have which results. And even when you need the math, you won't necessarily need to show it in any end result. You can just translate what the equation is saying into plain English and publish the latter. And the math you'll need is almost never as complicated as it is in the sciences, or even archaeology (where rather advanced Bayesian methods are already being deployed). It's usually very simple arithmetic, using basic *a fortiori* estimates. Thus its application in practice simply isn't difficult or time-consuming enough to oppose its use in history. To the contrary, as both a method in itself and a practical epistemological tool, its utility in history is considerable. The following chapters will be devoted to establishing that.

Canon of probabilities

To ensure the greatest consistency and the least contention, we can conform our inputs to a table of just eleven probabilities representing common historical judgments, which you will routinely find spelled out in English with the same or equivalent wording throughout professional literature in the field.[34] Only where and when we can convincingly demonstrate a probability more precise need we deviate from the following canon:

"Virtually Impossible" = 0.0001% (1 in 1,000,000) = 0.000001
"Extremely Improbable" = 1% (1 in 100) = 0.01
"Very Improbable" = 5% (1 in 20) = 0.05
"Improbable" = 20% (1 in 5) = 0.20
"Slightly Improbable" = 40% (2 in 5) = 0.40
"Even Odds" = 50% (50/50) = 0.50
"Slightly Probable" = 60% (2 in 5 chance of being otherwise) = 0.60

"Probable" = 80% (only 1 in 5 chance of being otherwise) = 0.80

"Very Probable" = 95% (only 1 in 20 chance of being otherwise) = 0.95

"Extremely Probable" = 99% (only 1 in 100 chance of being otherwise) = 0.99

"Virtually Certain" = 99.9999% (or 1 in 1,000,000 otherwise) = 0.999999

Obviously, you will often be able to argue for other values between these, or values vastly lower or higher than the extremes given. But unless you can be more precise, you should only employ the values here (or from some comparable canon) that support your conclusion *a fortiori* so that any adjustments in the direction of the correct values will only make your conclusions stronger.

Of course, we can always apply other canons. For instance, my assignment of the phrase "extremely improbable" to one in one hundred odds may fit how historians commonly speak, but not other contexts. No one would get into a car that had a one in one hundred chance of exploding, so we usually wouldn't say that that outcome is "extremely improbable." An extremely probable conclusion in history is therefore typically not anything anyone would bet his life on. How we decide what someone "means" when they use nonmathematical phrases to express probability is therefore context-dependent. Thus it's useful to look for measurable benchmarks within the same context and extrapolate from there. Before saying "I'd bet my life on it," what improbability of a car's exploding would you deem acceptable before getting into it?—and you can't say that probability must be zero, because it never is.

You can translate all ordinal benchmarks of confidence into probabilities by measuring them against known odds like this. If the odds were one in one hundred that a car you got into would explode, how "confident" would you be in that car's safety? Answer that and you've just ascertained what probability you mean by that expression of confidence. Likewise the other way around. If while on a basketball court you are only slightly confident you can make a hoop shot from where you are standing, how often does that mean you would make it if you tried it a dozen times? Once or twice? You've just ascertained that "slightly confident" means, to you, about one in six odds, or 17 percent. Or do you mean a little more than half the time, like maybe seven shots out of twelve? You've just ascertained that "slightly confident" means, to you, about seven in twelve odds, or 58 percent.

You might mean different things in different contexts, but then all you need do is find comparable benchmarks within each context, in order to build another translation key. All your nonmathematical expressions will still cover a range of probabilities, not just a single probability. But explore the limits on either side (e.g., how many hoop shots out of twenty, or a hundred) and you might find that to you "slightly confident" never means less than 51 percent or more than 66 percent, so it therefore means "between 51 and 66 percent," inclusively); or you might find you'd be "very confident" in a car's safety only if it exploded no more than once in a billion times, but you'd be just about as confident if if were once in a trillion, so "very confident" means to you "between one in a million and one in a trillion." And so on. You can then pick the limit that corresponds to an *a fortiori* probability. This works even if you are translating phrases with words other than "confident," like "sure," "certain," "credible," "likely," "believable," or anything else. How "believable" is it, for instance, that in a friendly game of basketball you will make a regulation free throw? That will certainly be a function of how often you make them and how often you don't. Betting odds follow, and from that, a probability.

FROM BAYES'S THEOREM TO HISTORICAL METHOD

This chapter hopefully provided an adequate primer to Bayes's Theorem, at least as far as applying it to typical cases of historical argument. But one might still ask, does BT *actually* underlie all valid historical methods? Aren't there other valid methods that work better? To that question we now turn.

CHAPTER 4
BAYESIAN ANALYSIS OF HISTORICAL METHODS

As one analyst put it, BT actually explains "what are regarded as sound methodological procedures" and reveals "the infirmities of what are acknowledged as unsound procedures" in almost any empirical field.[1] In other words, Bayes's Theorem underlies all other methodologies and thus explains why certain methods are regarded as sound, and others not—even when advocates or detractors of various methods are unaware of BT's capability in this regard. This entails a testable prediction, that all valid empirical methods reduce to BT: any method you propose will either be logically invalid or it will be described by BT. One might challenge how universally that's true, but here I will demonstrate that it at least holds for historical methods. I'll start with the most widely applicable examples, increasing in degrees of generalization, then test a few common methods of narrower scope.

Years ago I described two historical methods as defining the best that historians have deployed: the **Argument to the Best Explanation (ABE)** and the **Argument from Evidence (AFE)**.[2] The literature on historical method and the epistemology of history essentially supports this conclusion, all of it being reducible to one or the other.[3] Yet I have since discovered that everything I had argued can be better framed in Bayesian terms, especially since neither the ABE nor the AFE solves the problems I now identify as plaguing every historical method: establishing *logical validity* and *epistemic sufficiency*; in other words, why should the conclusion of these arguments be deemed logically valid and when is the evidence enough to warrant belief in the conclusion?).

THE ARGUMENT FROM EVIDENCE (AFE)

According to the AFE, there are at least five respected categories of historical evidence, and the more a conclusion has support from each category, the more likely it is to be true. Those categories are:

Physical-Historical Necessity: the degree to which history could not have proceeded as it did had the event(s) not occurred; that is, the degree to which the event is required to account for all subsequent history.

Direct Physical Evidence: archaeological evidence, material evidence, evidence physically produced by the event(s) or person(s) in question.

Unbiased or Counterbiased Corroboration: witnesses who have no known motive to lie or exaggerate, or even a motive to lie or exaggerate in the opposite direction.

Credible Critical Accounts: accounts by known scholars of the period that exhibit the use of a critical analysis and evaluation of multiple lines of evidence (as opposed to just repeating a story).

Eyewitness Accounts: accounts by actual eyewitnesses of the event(s) or person(s) in question.

All these amount to saying "these categories of evidence are unlikely to exist unless the proposed event happened," which in Bayesian terms means a low $P(e|\sim h.b)$ relative to a high $P(e|h.b)$. That's how you represent the fact in BT that the evidence (e.g., subsequent history, eyewitness testimony, etc.) is very improbable unless h (the hypothesized event[s]) happened. For example, Julius Caesar's conclusive capture of Rome and his unchallenged firsthand account of crossing the Rubicon would both have been improbable unless Caesar actually crossed the Rubicon. Not impossible, but improbable—certainly less probable than if Caesar had indeed done that. Which is simply a colloquial way of saying $P(e|\sim h.b)$ is low (because any plausible alternative account of things would render that evidence unlikely) and P(e|h.b) is high (as the event having occurred makes all that evidence very likely indeed). The five categories of evidence in the AFE just represent five different ways evidence can be more probable on *h* than ~*h*. And that list of five isn't even exhaustive. Though additions to it are likely to have much less weight (these five being typically the strongest types of evidence to have), there still are other kinds of evidence. Hence, BT is more generic and thus more universal, and all the premises of the AFE are logically included in the premises of BT.

This is confirmed by observing the effect of taking evidence away. If the physical-historical necessity of an event is minimal, then given solely that factor P(e|~h.b) and P(e|h.b) are about equal, because then the evidence (of the subsequent course of history) is as likely to have occurred even if our hypothesis is false. We thus need other evidence. Indeed, if we should expect the subsequent course of history to have been *different* on *h*, then the value of that evidence is actually reversed and counts *against* our hypothesis. Then it is P(e|~h.b) that is high, and P(e|h.b) that is low. Likewise, if there is no physical evidence and if that absence of evidence is just as likely on either *h* or ~*h* (such as due to the rarity of evidence surviving, as is frequently the case in ancient history), then again the consequents are equal (or near enough—see my discussion of 'evidence loss' in chapter 6, page 219). But if that absence of evidence is unusual on *h* (as is often the case in more modern history, where we actually expect physical and documentary evidence), then its absence argues against *h*, and P(e|~h.b) is then higher than P(e|h.b), sometimes very much higher (I'll examine the specific logic of an Argument from Silence later in this chapter, page 117). The same reasoning can be followed through for the other three categories of evidence. In fact, you can use BT to fully analyze the consequences of differing *degrees* of evidence. For example, just how unbiased or counterbiased a source is may not be so black and white. If it is only slightly unbiased, then the degree by which it will lower P(e|~h.b) will be smaller than if it were an ideally neutral source (such as someone who doesn't even understand the significance of what they are attesting to).[4]

In every case, the degree to which the consequents differ from each other will reflect the degree to which the evidence is unexpected on either *h* or ~*h*.

It should not be difficult to select an *a fortiori* measure of degree from the canon of probabilities supplied in chapter 3 (and repeated in the appendix, page 286). If you want to be sure whether *h* is credible, and the evidence *does* entail P(e|h.b) is higher than P(e|~h.b), then select a measure of degree that is less than you are certain it is. For example, if you are sure the odds must be way lower than one in one hundred that all this evidence would exist if ~*h*, then select one in one hundred (0.01), or even one in twenty (0.05), as reflecting the degree of difference between the consequents. For example, you might assign P(e|h.b) = 1 and P(e|~h.b) = 0.05. Then whatever result you get, if it supports believing *h*, it will support that conclusion even more, since you know any correction of your estimated probabilities can only *raise* the final epistemic probability that *h* is true (as explained in chapter 3, page 85).

What about prior probability? The AFE as stated implicitly assumes all competing hypotheses are equally likely prior to considering the evidence for them—which limits its validity. For example, the famous rain miracle of Marcus Aurelius *only* has extant evidence supporting magic or miracle as its explanation and description.[5] To what extent are we obliged to conclude that those reports are correct, rather than preferring a hypothesis of what happened that explains these reports *and* the event without recourse to sorcery or meddling gods? The AFE can't answer that question in any logically valid way. But it can answer the question of whether the evidence we have is very likely or unlikely on any given hypothesis, which corresponds to the consequent probabilities in BT. So when the priors are indeed equal, or close enough that the evidence-produced disparity in consequents would easily overwhelm them, the AFE gives intuitively correct results. Thus, BT formally represents the logic of the AFE, and the AFE is only valid insofar as it can be validly represented with BT.

THE ARGUMENT TO THE BEST EXPLANATION (ABE)

According to the ABE, there are five qualities a *theory* can possess in respect to the evidence, such that the more it fulfills those qualities over any alternative explanation of the same evidence, the more likely it is to be true. Those qualities are (quoting and paraphrasing McCullagh[6]):

Plausibility: the hypothesis must conform to the expectations set by our background knowledge; formally, "it must be implied . . . by a greater variety of accepted truths than any other, and be implied more strongly than any other; and its probable negation must be implied by fewer beliefs, and implied less strongly than any other."

Ad Hocness: the hypothesis must rely on the fewest *ad hoc* assumptions possible to explain the evidence, that is, assumptions for which there is no evidence or established agreement, or things just made up to force the hypothesis to fit; formally, "it must include fewer new suppositions about the past," and about the nature of man and the world, "which are not already implied to some extent by existing beliefs."

Explanatory Power: the hypothesis must make the evidence we have very probable; formally, "it must make the observation statements it implies more probable than any other."

Explanatory Fitness: the hypothesis must not contradict any evidence or well-established beliefs, or at least contradict them much less than any competing theory does (since contradictory evidence can be explained away by various devices, sometimes legitimately; indeed a new result contradicting a prior belief is exactly how we discover a prior belief is false); formally, "when conjoined with accepted truths it must imply fewer observation statements . . . which are believed to be false."

Explanatory Scope: the hypothesis must explain more of the evidence we have than any other hypothesis can; formally, "it must imply a greater variety of observation statements" that can be checked against surviving evidence.

This list is essentially just a lay summary of BT. For each criterion, the question hinges on how much h exceeds all the alternatives (which together constitute $\sim h$), which requires a measure of degree, which is by definition mathematical. And such a measure of degree (by which h exceeds $\sim h$) is exactly what BT employs. With the ABE, the end result requires combining all five factors, which can be complicated if competing hypotheses match or exceed each other on different criteria. Yet the ABE provides no means of ascertaining the effect of combining all five criteria. Even in straightforward cases, where h exceeds $\sim h$ on every criterion, what degree of belief is warranted by that degree of superiority on each of those five criteria? The ABE alone cannot answer that question. BT can; likewise on complex cases. Hence, BT is superior to the ABE.

In fact, the ABE criteria themselves are just colloquial versions of the premises in BT. **Plausibility** combined with **Ad Hocness** is simply a description of prior probability. Hence, prior probability in BT *is* the combination of 'Plausibility' and 'Ad Hocness' in the ABE. The more our background evidence renders our theory more typical, the higher its prior. The less typical our background evidence renders our theory, the lower its prior. And requiring new suppositions (about the world and the past) entails an untypical explanation, whereas an explanation fully (and indeed better) supported by background knowledge is thereby, by definition, more typical. BT represents this fact by a difference in priors. And this can only be validly represented when all possible explanations are represented in relative terms to each other, which only a proper application of BT ensures (as in BT the sum of all priors must equal one). Adding *ad hoc* elements likewise includes the tactic of inventing 'excuses' for the evidence not fitting your theory, which also lowers prior probability (as I demonstrated in chapter 3, page 80), which means that *not* depending on such excuses necessarily *raises* the prior (by exactly as much as depending on them would have lowered it).

The last three criteria then combine to entail the consequent probabilities in BT. **Explanatory Power** is almost an exact description of consequent probability. It only lacks reference to the circumstance of a hypothesis entailing a low consequent—which is accomplished by the Criterion of **Explanatory Fitness**. In fact, those two criteria are obviously two sides of the same measure: one refers to evidence that is very expected on a hypothesis, the other to evidence that is very unexpected on that same hypothesis; the one entailing a high consequent, the other a low one. The only thing missing is the middle possibility: evidence some hypotheses neither predict nor contradict. And that's essentially what **Explanatory Scope** picks up, by addressing facts a theory makes likely but that another theory makes neither likely nor unlikely. Combine all three, along with the fact that it is stated in each that you are measuring the degree of difference between the tested hypothesis and all its alternatives, and you simply have the difference of consequent probabilities measured in BT.

To see why they're the same, once again we only need examine what happens when evidence is added or taken away. Increasing Explanatory Scope (relative to a competing hypothesis) entails decreasing $P(e|\sim h.b)$, since not explaining those facts renders them less probable, while explaining them renders them probable, keeping $P(e|h.b)$ high. In contrast, increasing Explanatory Fitness or Power directly entails increasing $P(e|h.b)$, while decreasing either of them entails decreasing it. And insofar as a competing hypothesis itself has a high or low Fitness or Power, $P(e|\sim h.b)$ is also rendered high or low accordingly; likewise if $\sim h$ has the greater Scope, then it's $P(e|h.b)$ that drops instead of rises. Thus, BT describes and in fact legitimizes the ABE. And yet BT is superior to the ABE, by having the precision that guarantees the logical clarity and validity of its results and ensures that ambiguous and unstated assumptions (about measures of relative degree) become clear and stated, and thus open to challenge and thus requiring sounder defense, thereby also ensuring its premises will be more sound than is likely under the vague structure of the ABE.

Thus, BT achieves greater soundness and validity than the ABE. Which reminds us again that any argument against the applicability of BT to history logically entails the same argument against the applicability of the ABE to history, and the AFE as well; and, I predict, all methodologies whatever, insofar as they have any validity to begin with.

THE HYPOTHETICO-DEDUCTIVE METHOD (HDM)

In fact, *any* valid form of hypothetico-deductive method is described by BT. Even the principle of Ockham's Razor (when validly formulated) follows necessarily from BT.[7] **Hypothetico-Deductive Method** (HDM) is the procedure of forming a hypothesis *h*, deducing observations that would be made if *h* is true and observations that would be made if *h* is false, and making many observations until the probability of *h* either far exceeds all known alternatives, or drops below all credibility. This exactly describes the BT ratio of consequents (or else the iterative use of BT, on which see chapter 5, page 168). But you can always redesign any alternative hypothesis so that ~*h* also predicts all the same observations as *h*. So how do you tell the difference? In practice, if your predictions are good ones (i.e., the evidence that results from many diverse observations normally entails a high ratio of consequents favoring *h*), then redesigning a new hypothesis so as to "just happen" to make all those same predictions will require an enormous Rube Goldbergesque contraption of additional assumptions, whereas the tested hypothesis is simple and requires few new assumptions; at which point, Ockham's Razor is invoked, and *h* thus prevails over ~*h*.

This difference between such explanations is described by the logic of prior probability in BT.[8] Thus, Ockham's Razor is merely a declaration of that fact: the more assumptions you tack onto any *h* (especially novel assumptions), the lower its prior must be (as I demonstrated in chapter 3, page 80). Because any collection of "coincidences" like that is less typical, and thus less probable, than causes that do not require them (much less so many of them). This follows not merely from the addition of novel assumptions, but also the addition of improbable assumptions. For example, that the CIA might meddle with science experiments is not impossible (in fact such meddling in other affairs is actually attested, as are some of the means and motives necessary to meddle in the same way with scientific research), yet any *h* that *relied* on the unverified assumption that the CIA was meddling with your experiment would by virtue of that fact be far less probable than almost any *h* excluding that assumption, and thus 'CIA interference' is usually (and rightly) axed by Ockham's Razor. And that represents the lowering of priors based on our background knowledge regarding what is and isn't typical (the CIA meddling in science experiments isn't typical).

Thus all scientific methods, which are simply iterations of HDM, are described by BT.[9] Historical methods are identical to scientific methods in this respect, being just another set of iterations of HDM. In fact, many sciences *are* historical, for example, geology, cosmology, paleontology, criminal forensics, all of which explore not merely scientific generalizations but historical particulars, such as when the Big Bang occurred, or how the solar system formed, or exactly when or where a large asteroid struck the earth, or when a volcano erupted and what resulted from it, or what happened to a specific species in a specific historical period, or who committed what crime when. Not even the claim that historians must deal with human thoughts and intentions makes a difference, as these are as much a necessary occupation of psychologists, economists, sociologists, and anthropologists. It's also fundamental to the scientific study of game theory and all of cognitive science. Nor is there any demarcation based on the role of controlled experiments. Much of science does not rely on experiments but primarily involves field observations (e.g., astronomy, zoology, ecology, paleontology), an approach to evidence directly analogous to the historian (most clearly parallel in the science of archaeology, but "field observations" of the artifacts we call "texts" and "documents" is just as analogous). Conversely, experiments sometimes do have a place in historical methodology.[10] And as noted in chapter 3 (page 47), science is actually dependent on history, just as much as history depends on science.

So there is no qualitative difference. History is thus continuous with science. The difference between them is only quantitative: history must work with much less data, of much less reliability. Therefore, its results have less certainty and less precision. BT even explains this: the data available to science are of such scale and quality as to raise the final epistemic probability of its results to incredible heights (so high, in fact, often no one even bothers to calculate them, nor need they). But in history we are almost always dealing with final probabilities that, however high they may be, nevertheless allow a possibility of being mistaken that isn't negligible—to such a degree, in fact, that scientists would reject comparably uncertain results. But their rejecting such results does not mean those results are not believable, only that they do not obtain a scientific degree of certainty. The demarcation between science and nonscience is not the demarcation between believable and unbelievable conclusions. It is merely the demarcation between conclusions that are, for all intents and purposes, decisively certain (albeit still revisable), and conclusions that are not. But many conclusions can be believed with legitimate confidence without being decisively certain. We meet with such beliefs routinely in journalism, economics, and daily life. So as long as we face this fact and accept history is like that, we can proceed scientifically without pretending to the certainty of scientific results.

FORMAL PROOF OF UNIVERSAL APPLICABILITY

Since BT fully describes HDM without remainder, and HDM is a higher-level generalization of all historical methods, including the AFE and ABE, we could simply conclude here and now that Bayes's Theorem models and describes all valid historical methods. No other method is needed, apart from the endless plethora of techniques that will be required to apply BT to specific cases—of which the AFE and ABE represent highly generalized examples, but examples at even lower levels of generalization could be explored as well (such as the methods of textual criticism, demographics, or stylometrics). All become logically valid only insofar as they conform to BT and thus are better informed when carried out with full awareness of their Bayesian underpinning.

This should already be sufficiently clear by now, but there are always naysayers. For them, I shall establish this conclusion by formal logic.

P1. BT is a logically proven theorem.
P2. No argument is valid that contradicts a logically proven theorem.
C1. Therefore, no argument is valid that contradicts BT.

P1 is an established fact (see note 9 for chapter 3, page 300). P2 is true by definition, that is, what it is to be a logically valid form of argument is to be consistent with formal logic, and all logical proofs are consistent with formal logic, ergo, to be inconsistent with a logical proof is to be inconsistent with formal logic, which entails by definition an invalid argument. Formally, if B = 'we must accept the sound conclusions of formal logic,' A = 'BT is true,' and C = 'there is some historical method that is logically valid but contradicts BT,' then:

P3. If B, then A.
P4. If C, then ~A.
C2. Therefore, if B, then ~C.

This is a logically necessary truth.[11] Therefore there can be no valid historical method that contradicts BT. This leaves only two other possibilities: either (a) all valid historical methods are fully modeled and described by BT (and are thereby reducible to BT), or (b) there is at least one valid historical method that does not contradict BT but that nevertheless entails a different epistemic probability than BT for at least one historical claim h. The only way that can be logically possible is if there is something that could be said about the epistemic probability of h that is not said about the epistemic probability of h in BT. Because if BT already says that about h (i.e., if it already contains a premise about the effect of that same fact on the epistemic probability of h), then the only way any method can say anything different (about the effect of that fact on the epistemic probability of h) is by contradicting BT, which we just demonstrated is logically impossible. Any method that did that would have to be logically invalid.

Can that point be proven? Yes. About the epistemic probability of h a method can say all the same things as BT (in which case it must give the same conclusion as BT, or else it is contradicting BT and therefore contradicting logic), or it can say more things than BT, or it can say fewer things than BT. Methods that say less than BT, yet declare an epistemic probability for h, can only be methods that ignore known facts that affect the epistemic probability of h, and such methods are necessarily invalid and must thereby be excluded, which leaves only methods that say more than BT. This does not mean, however, all methods are invalid that only *seem* to say less than BT but that in fact (implicitly) say all the same things as BT. Just as the AFE is logically valid only if it is allowed to implicitly assume all prior probabilities are equal, so, too, *any* valid statistical arguments that ignore considerations of prior probability are still implicitly assuming *some* prior probability, whereas if they are not assuming that, then they are logically invalid.[12] For not assuming the priors are equal entails assuming the priors must be different, and any method that assumes the prior probability of *h* is different from ~*h* but does not enter that difference *somehow* into its calculation of the final probability of *h* is willfully illogical—at any rate, it thereby directly contradicts BT, which, as proved, is contrary to logic.[13] So all that's left are methods that say *more* than BT. But we know of nothing that can be said that would validly affect the epistemic probability of *h* other than what is already said by the premises in BT. And if a method says nothing different about the probability of a claim being true than is already said by BT, that method can be fully replaced by BT without logical consequence.

Accordingly, I propose the following testable hypothesis:

> **P5.** Anything that can be said about any historical claim *h* that makes any valid difference to the probability that *h* is true will either (a) make *h* more or less likely on considerations of background knowledge alone or (b) make the evidence more or less likely on considerations of the deductive predictions of *h* given that same background knowledge or (c) make the evidence more or less likely on considerations of the deductive predictions of some other claim (a claim which entails *h* is false) given that same background knowledge.

Thus to reject my conclusion (that all valid historical methods are reducible to BT) requires providing a counter-example to P5—which I predict no one can do. It's probably impossible, as by definition *b* and *e* encompass all data (i.e., the union of those two sets produces the set of all things known), and *h* and ~*h* encompass all theories, and BT logically includes every probability entailed by *b* and *e* on every theory. For example, P(h|b.e) is transitively identical to P(h|e.b) and is by definition the probability that *h* is true given all available knowledge (the union of *e* and *b*), so there is by definition *no other knowledge that can alter that probability.* Yet P(h|e.b) follows by logical necessity from P(h|b), P(e|h.b), and P(e|~h.b); therefore no other probability can have any relevance to determining P(h|e.b).

That means the following is true by definition:

P6. Making *h* more or less likely on considerations of background knowledge alone is the premise P(h|b) in BT; making the evidence more or less likely on considerations of the deductive predictions of *h* on that same background knowledge is the premise P(e|h.b) in BT; making the evidence more or less likely on considerations of the deductive predictions of some other claim that entails *h* is false is the premise P(e|~h.b) in BT; any value for P(h|b) entails the value for the premise P(~h|b) in BT; and these exhaust all the premises in BT.

Formally, if C = 'a valid historical method that contradicts BT,' D = 'a valid historical method fully modeled and described by (and thereby reducible to) BT,' and E = 'a valid historical method that is consistent with but only partly modeled and described by BT,' then:

P8. Either C, D, or E. (proper trichotomy)
P9. ~C. (from C2.)
C3. Therefore, either D or E.

P10. If P5 and P6, then ~E.
P11. P5 and P6.
C4. Therefore, ~E.

P12. If ~C and ~E, then only D. (from P8.)
P13. ~C and ~E. (from C2 and C4.)
C5. Therefore, only D.

Therefore, only a valid historical method fully modeled and described by (and thereby reducible to) BT exists. In other words, no other valid historical methods exist.[14]

Indeed, I believe that for any claim h (whether in history or any other subject of knowledge whatever), if h is capable of being true or false, then its probability of being true (given what you happen to know at the time) is exactly and only the probability entailed by BT. Therefore, no one can reject the valid and sound conclusions of BT in any subject of factual inquiry, and no one can claim that BT does not or cannot determine the probability that any h is true (given all present knowledge). The preceding proof already entails this must be true for claims about history, and that is sufficient for my present purpose. But I shall pause to demonstrate the broader thesis as well, as it reinforces the narrower.

The broader thesis follows from an argument I developed in chapter three (pages 83–88). Of the four probabilities in BT, one entails another (each prior is always the converse of the other), so there are only three independent statements of probability in BT. For each of those you either know that its value is higher than 0.5 or lower than 0.5, or else so far as you know it *is* 0.5—because if you don't know whether it's higher or lower, then by definition so far as you know it's as likely as not. If it weren't as likely as not, then by definition that would mean you know it is not 0.5, *which entails you know it is either higher or lower*. Therefore, for every premise in BT, you always know its probability is either A (0.5), B (higher than 0.5), or C (lower than 0.5), "so far as you know," and that exhausts all logical possibilities. But no matter what value thereby results for each and every premise in BT (whether A, B, or C), a conclusion necessarily then follows regarding the probability that h is true, given what you are thus claiming to know—even if that probability is flat out 50/50. For example, for any claim h, if you know nothing about what any of these three probabilities for it are, then so far as you know they are all 0.5, which logically entails the posterior epistemic probability, the probability that h is true ("given what you know so far") is 0.5. And because BT is logically valid, you must always accept its conclusions when you accept its premises. Thus the only way to deny such a conclusion is to affirm different premises (e.g., that one of those three probabilities is *not* 0.5), in other words, to affirm that you know what those probabilities are (at least well enough to know they aren't the probabilities just affirmed).

Of course by arguing *a fortiori* you might say something like it's "0.5 *or higher,*" but even in that case mathematically you are entering a 0.5 in the equation, so it amounts to affirming that the probability is simply 0.5. What you then do with the conclusion is determined by whether you think the "true" probability could be higher or could be lower (the method of arguing *a fortiori*, which I discussed in chapter 3, page 85). But mathematically you still have to choose to set the limit of your margin of error at 0.5, or above 0.5, or below 0.5. There is no fourth alternative. That some conclusion is then always entailed I demonstrate with a logical flow chart in the appendix (page 286), where BT conclusions are shown to follow even from the simplest trichotomy here proposed, that is, that each premise has a value of either 0.5 or < 0.5 or > 0.5. But the same analysis follows for any degree of precision your knowledge can honestly claim.

For example, if you must admit P(h|b) is > 0.5, then for any arbitrary number above 0.5, for instance $n = 0.75$, either you believe (A) P(h|b) is n, or (B) P(h|b) is greater than n, or (C) P(h|b) is less than n, or else (D) you can't claim to know anything more than that n is above 0.5. And so on, for any other n. You cannot deny any of these possibilities without affirming one of them, and once one of them is affirmed, a conclusion always necessarily follows from BT as to what the probability is that h is true given what you know. It might still be that *or higher*, or that *or lower*, depending on which limit of your margin of error you are defining, but that's still a probability that cannot be rejected without giving in and accepting BT and just affirming different values for its premises.

To be a little more specific, if you affirm "very high confidence" that h would turn out to be true before looking at the specific evidence for h, then you *cannot logically believe* P(h|b) is less than 0.75. Because to simultaneously assert "I have a very high confidence that h will turn out to be true" and "I believe h will be true only three out of every four times" is to hold two contradictory beliefs. To believe h will be true only three out of four times is simply not to believe "with very high confidence" that h will turn out to be true. That would be a confidence only *somewhat* high. Just ask yourself again whether you'd get into a car that had a one in four chance of exploding, and then ask why you would still claim to have a "very high confidence" in the safety of that car. Or if that context is too extreme, ask yourself if you'd bet your career on a result that had a one in four chance of soon being refuted, and then ask why you'd have a "very high confidence" in a result like that. This becomes all the clearer as your certainty increases. If you are certain h will turn out to be false, because you rightly believe it is wildly improbable for h to be true (e.g., as when h = "Caesar rode a winged horse and camped on the moon"), then you cannot believe that P(h|b) is even as high as 0.01 (or 1 percent), much less any higher. This was already evident in chapter 3's opening example of the sun being supernaturally eclipsed for three hours.

It follows that no other method of inference, such as using ordinal or qualitative rankings of confidence absent any reference to probabilities, can supplant BT. It can work alongside it, as a heuristic or simplified way of getting the same result. But you can't get a valid conclusion from any other method and have that conclusion contradict BT. And for every hypothesis, BT entails some conclusion (for whatever state of knowledge you are presently in). Because for any probability in BT (whether the prior or either consequent), you will always have some confident belief that it is at least some value or higher, or some value or lower, and this belief logically requires you to accept the conclusions that follow from that belief, in the very manner BT entails.

Again, for example, if you cannot deny that the prior probability that the sun went out for three hours is "not higher than 0.01," and you cannot deny that the consequent probability that no one else would notice this is "not higher than 0.01," and you cannot deny that the consequent probability is anything significantly less than 1 that *some* sacred writing about Jesus would claim the sun went out *if that claim was fabricated*, then you simply cannot deny that the epistemic probability that the sun went out is not higher than 0.001 / (0.001 + 0.99) = 0.00101, or roughly one tenth of 1 percent (and is likely much lower). You are logically obligated to agree that this is true, *unless and until you can demonstrate any of those underlying probabilities to be different*. Resorting to other methods of inference simply cannot extricate you from this obligation. At best they can only confirm the same result, and thus simply corroborate BT.

My conclusion is therefore inescapable. For each of the three premises in BT (the prior and the two consequents), for any claim *h*, always there is some probability that you will be confident declaring, such that any amount beyond it you would not be confident declaring. That probability is then the only one that will entail a conclusion you can be confident in—because to get a different conclusion out of BT, you must input a different probability, yet if you are inputting a probability you are not confident in, then you cannot be confident in whatever conclusion that that probability entails. Because every weakness in an argument's premises always translates to the conclusion. Thus a conclusion that can only be arrived at by affirming premises you are not confident are true is by definition a conclusion you are not confident is true. But for every probability you are confident in, BT entails a conclusion you are logically compelled to be *equally confident in*.

There is no way around this. Because no matter how ignorant you claim to be, some value always necessarily follows for $P(h|e.b)$, since by definition that is a probability conditional on what you know, and therefore a probability follows even if what you know is nothing (that probability would always be 0.5). So there is always a Bayesian probability that *h* is true given everything you know. This is due to the key difference between epistemic and physical probabilities as demarcated in chapter 2 (page 24). With regard to physical probabilities, you can legitimately say you simply have no idea what the probability is (and therefore you are not obligated to pick any one of the only logical possibilities available), but you cannot legitimately say this with regard to epistemic probabilities. To say you have no idea, before looking at evidence *e*, whether *h* or ~*h* is true logically entails that for you (i.e., so far as you presently know), *h* is as likely as ~*h* (until you see some evidence). Because the only way you can claim to know that *h* is *not* as likely as ~*h* is to claim to know that *h* is more likely than ~*h* or that *h* is less likely than ~*h*; but you have already affirmed that you do not know either. Therefore, so far as you know, *h* is as likely as ~*h*. Which when translated into mathematical notation simply means $P(h|b)$ = 0.5. So even denying there is a probability always entails there is a probability—for you, and given the information you have at that point in time.

It follows that you are always in some state of knowledge or ignorance that entails an epistemic probability for any *h* according to BT, and no other method of inference can validly contradict it. BT therefore underlies all valid empirical reasoning. Or so I contend.

Even if you disagree with that broader thesis, you still cannot deny that BT underlies all valid *historical* reasoning, as that at least I formally proved earlier. Applying this knowledge now allows us to test the validity of any methodological principle in the study of history. Two major examples are the 'Argument from Silence' and what's amusingly called the 'Smell Test.' The following analysis of these can serve as a model by which to evaluate any other methodological principle in history.

BAYESIAN ANALYSIS OF THE 'SMELL TEST'

The 'Smell Test' is a common methodological principle in the study of myth, legend, and hagiography. This test can be most simply stated as "if it sounds unbelievable, it probably is." When we hear tales of talking dogs and flying wizards, we don't take them seriously, even for a moment. We immediately rule them out as fabrications. We usually don't investigate. We don't wait until we can find evidence against the claim. We know right from the start the tale is bogus. Yet the only basis for this judgment is the Smell Test. Is that test valid?

It is certainly ubiquitously accepted by historians in every field. It is suspiciously *only* rejected by religious believers, and then only when it's applied to amazing claims they prefer to believe. They ground this rejection in the claim that we shouldn't be biased against the supernatural, and God can do anything. Yet if they *honestly* believed in those principles they would be compelled to concede the miracle claims of *every* religion "because you shouldn't be biased against the supernatural, and God can do anything." This includes all the pagan miracles (incredible apparitions of goddesses, mass resurrections of cooked fish, wondrous healings, and teleportations), Muslim miracles (splitting moons, wailing trees, flights to outer space), Buddhist miracles (bilocation, levitation, creating golden ladders with a mere thought), and indeed every and any amazing claim whatever. Tales "proving" reincarnation? We can't reject them—because God can do anything. Ghosts confirming to the living that heaven is run by a Chinese magnate and his staff? We can't rule it out. That would be bias against the supernatural.

Honestly living that way would be impossible. You would have to believe everything you read or hear unless you can specifically present evidence sufficient to discount it: an impossible task. You would be left with a belief system hopelessly frightening and contradictory—and mired in a thousand false beliefs. Such behavior also goes against all established background knowledge, which contains endless examples of miracle claims refuted by fortuitous inquiry (and no good case of any miracle claim *surviving* such inquiry).[15] In other words, our bias against the supernatural is *warranted*, just as our bias against the honesty of politicians is warranted: we've caught them being dishonest so many times it would be foolish to implicitly trust *anyone* in politics. Likewise, amazing tales: we've caught them being fabricated so many times it would be foolish to implicitly trust *any* of them.

The Smell Test thus represents an intuitive recognition of: (a) the low prior probability of the events described (i.e., $P(h|b) \ll 0.5$); (b) the ease with which the evidence could be fabricated (i.e., $P(e|{\sim}h.b)$ is always high, unless we have sufficient evidence to the contrary), in fact often the ease with which such an event *if real* would produce or entail much better evidence (i.e., $P(e|h.b)$ is often low); (c) how typically miracle claims are deliberately positioned in places and times where a reliable verification is impossible (and when such verification *is* possible, are refuted), which fact alone makes them *all* inherently suspicious; and (d) sometimes the similarity of a miracle story to other tales told in the same time and culture is additionally suspect, like the odd frequency with which gods in the ancient West rose from the dead, transformed water into wine, or resurrected dead fish, oddities that curiously never occur anymore, and which are so culturally specific as to suggest more obvious origins in storytelling.[16]

Both (c) and (d) can raise the consequent probability of nonmiraculous explanations, and also reduce the consequent probability of the miraculous.

But condition (a) is the point just made: such claims are contrary to reality as we know it. This doesn't just mean only miracles, but all wonders, like implausible coincidences and unrealistic social reactions and behaviors. Hence, the issue is not the presumption that miracles never happen, but the documented fact that, if they happen, they happen exceedingly rarely (just as implausible coincidences and unrealistic human behaviors do), whereas false tales of the fantastic happen with exceeding frequency; and likewise, the fact that miracles suspiciously happen all the time only in historical periods (or geographical regions) that are comparatively illiterate, super-stitious, or unenlightened, in conditions lacking the means of verifying no shenanigans were involved (in either the event or its telling), whereas in ages and places where we have widespread education and organized skep-ticism and the tools and opportunity to test wild claims, the phenomena always disappear. Both are established facts in *b* (our background knowl-edge) and thus all our estimates of probability must be conditioned on these facts. Even if you are a firm believer in the miraculous, the facts remain the same: most wondrous claims (by far) are bogus. Your priors *must* reflect that, regardless of your worldview. And like the Tibetan peasant claim in chapter 3 (page 72), when we lack specific evidence to confirm a claim, or lack the means to verify it by reliable tests, the priors must dictate what is reasonable to believe. That reasoning is both logically valid and sound. Thus, BT confirms that the Smell Test is valid, even on point (a) alone.

Conditions (b), (c), and (d) only strengthen this conclusion. Even outside the context of wondrous claims, ancient texts are full of lies and falsehoods, even when generated by eyewitnesses, contemporaries, and critical historians, or anyone who ought to have known better.[17] Our back-ground knowledge also establishes how easy it is to rapidly fabricate and disseminate false stories, even without challenge (like the darkening of the sun with which we began in chapter 3; and more examples I'll explore in the next volume), *and* how easy it is for a claimed miracle to entail evi-dence we curiously don't have. The darkening of the sun predicts a vast quantity of evidence that, by not existing, disconfirms the story. Likewise, the frequency of resurrection stories in antiquity entails a phenomenon that should still be observed with the same frequency, yet is not (except in such mundane ways as to refute any miracle claimed to be analogous—such as from the application of CPR and ordinary cases of misdiagnosed death). Thus, the disappearance of this phenomenon is an unexpected piece of evidence on the theory that any resurrection is real, just as the disappearance of angels and gods who used to descend and deliver speeches with surprising frequency in antiquity is unexpected on the theory that these things ever really happened. It's not *impossible* that "things just changed," but it *is* improbable—because we cannot predict from any *established* theory that such a change would indeed have happened, much less happened conveniently as soon as we had better methods and means to test such claims (and it is precisely that *coincidence* that is otherwise very improbable). Any logically valid argument must take this improbability into account.[18] Thus, incredible claims can only pass the Smell Test if they have correspondingly strong evidence in their support, which means evidence that is even more improbable on the claim's being false than the claim's being true is already improbable on prior considerations ((a) through (d)). For example, if in all past cases a claim's being true is a tenth as likely as its being false, then to believe that claim we need evidence that's over ten times *more* unlikely on any other explanation. That is, if $P(\sim h|b) = 10 \times P(h|b)$, then $P(e|h.b)$ must exceed $10 \times P(e|\sim h.b)$ for h to be credible; and if $P(\sim h|b) = 1{,}000 \times P(h|b)$, then $P(e|h.b)$ must exceed $1{,}000 \times P(e|\sim h.b)$ for h to be credible; and so on. In other words, the Smell Test simply reduces to the principle "extraordinary claims require extraordinary evidence" (see chapter 3, page 72; chapter 5, page 177; and chapter 6, page 253). Which means the Smell Test reduces to BT.

BAYESIAN ANALYSIS OF THE ARGUMENT FROM SILENCE

Historians routinely rely on Arguments from Silence: when something *isn't* said or attested, we conclude it didn't happen. Such reasoning is often challenged with the quip "absence of evidence is not evidence of absence." But the truth is, absence of evidence *is* evidence of absence—but only when that evidence is expected. You also sometimes hear the axiom "you can't prove a negative," but that's also false. Negatives are often quite easy to prove and we prove them all the time. In fact, logically, every positive claim entails a converse negative claim, thus merely in the act of proving a positive we have always proven a negative; often a great number of them.

The question of whether Jesus existed, for example, would be decisively proven in the negative by the recovery of an authenticated letter signed by the Apostle Peter outright saying that Jesus was only a cosmic being whose sojourn on earth was merely a symbolic myth, and who was only known to anyone through mystical perception. And we could have had a great deal more evidence than that—as we do for the ahistoricity of Betty Crocker, for example. Hence, proving a negative *in principle* is no difficulty. The ahistoricity of Moses and Abraham and all the other patriarchs is now generally accepted by scholars the world over as an established fact, quite rightly, and yet without even need of such a smoking gun as a contemporary epistle declaring them a fiction.[19] But can we validly argue that if Jesus didn't exist we would have such a letter from Peter (or any such evidence), and therefore the fact that we don't argues against the notion? Unfortunately, no, because we have little reason to expect such evidence to have survived for us to now have it. Indeed, there would be no reason for anyone actually to say Jesus didn't walk the earth until someone started saying he *did*. If that only happened after Peter died, he won't ever have written a letter gainsaying it. Whether we can expect *someone* to have done so, however, is a question I must ask in the next volume.

For the present, our concern is with when an Argument from Silence is valid and sound—and when it is not. The logical conditions have already been correctly stated:

> To be valid, the argument from silence must fulfill two conditions: the writer whose silence is invoked in proof of the non-reality of an alleged fact, would certainly have known about it had it been a fact; [and] knowing it, he would under the circumstances certainly have made mention of it. When these two conditions are fulfilled, the argument from silence proves its point with moral certainty.[20]

That would be a slam-dunk case. But a relatively weaker deployment is possible, to the extent that either condition is less certain. So it may only be "somewhat certain" that the relevant authors knew the fact and would mention it, in which case this argument can produce only a "somewhat certain" conclusion. Generally speaking, based on the hypothesized fact itself, and in conjunction with everything we know on abundant, reliable evidence, should we *expect* to have evidence of that fact? If the answer is *yes*, and yet no such evidence appears, then an Argument from Silence is strong. If the answer is *no*, then it's weak. Not having more evidence of the sun going out (examined in chapter 3) is a strong Argument from Silence, but not having a letter from the first Apostles explicitly declaring Jesus a fiction is weak. The examples of Caesar shaving or playing dice with a hooker (examined in chapter 2) are weaker still, being exactly what historians have in mind when declaring absence of evidence is not evidence of absence. Yet as the sun case proves, that rule does not always apply.

Once again, BT describes the logic of this argument. If on h we should expect some evidence e_1 given b (all our background knowledge) and yet we don't have e_1, then the consequent probability of h must be reduced— by exactly as much as lacking e_1 is unlikely (because the absence of that evidence is a part of the full e that must be explained by h). This same rule operates on the consequent of $\sim h$ as well, if $\sim h$ entails evidence we don't have. The tricky bit is the effect b has on this estimate. Our background knowledge establishes a very low expectation for the survival of evidence from antiquity, particularly the kind of evidence we would expect if Jesus didn't exist (I shall discuss the more generic problem of 'lost evidence' in chapter 6, page 219). However, that same background knowledge establishes a rather high expectation that the evidence that *did* survive would possess certain characteristics, and some scholars have argued that the surviving evidence is of a different character entirely. In my next volume I will discuss this oddity and how it might be dealt with. The point to observe here is that the Argument from Silence is a commonly accepted historical tool, and is logically valid precisely and only when it conforms to BT.

What else will Bayes's Theorem teach us about the methods particularly used by Jesus scholars? To that we now turn.

CHAPTER 5

BAYESIAN ANALYSIS OF HISTORICITY CRITERIA

In chapter 1, I demonstrated the growing loss of faith in the methodology of Jesus studies, as well as the inevitable consequence of relying on those invalid methodologies; namely, the plethora of contradictory conclusions about Jesus, each as confidently asserted as the next. This overall approach, which has so dismally failed, consists in the development and application of 'historicity criteria.' Chief among the cited defects of this approach was a failure to solve the Threshold Problem. When is the evidence enough to warrant believing any conclusion? No discussions of these criteria have made any headway in answering this question. Yet the question is inherently mathematical in nature. Only Bayes's Theorem can answer it. In chapter 3, I explained why the Threshold Problem requires mathematical reasoning to answer and how BT accomplishes that. And in chapter 4, I demonstrated that all valid historical methods represent different applications of Bayesian reasoning and methods that violate BT are not valid. What happens when we apply that same analysis to the popular historicity criteria?

Many dozens of criteria have been proposed in the field of Jesus studies; some overlap or just re-label another criterion. But at least eighteen distinctive criteria can be identified:

Dissimilarity: if dissimilar to Judaism or the early church, it's probably true

Embarrassment: if it was embarrassing, it must be true

Coherence: if it coheres with other confirmed data, it's likely true

Multiple Attestation: if attested in more than one source, it's more likely true

Explanatory Credibility: if its being true better explains later traditions, it's true

Contextual Plausibility: must be plausible in Judeo-Greco-Roman context

Historical Plausibility: must cohere with a plausible historical reconstruction

Natural Probability: must cohere with natural science (etc.)

Oral Preservability: must be capable of surviving oral transmission

Crucifixion: must explain (or make sense of) why Jesus was crucified

Fabricatory Trend: mustn't match trends in fabrication or embellishment

Least Distinctiveness: the simpler version is the more historical

Vividness of Narration: the more vivid, the more historical

Textual Variance: the more invariable a tradition, the more historical

Greek Context: credible, if context suggests parties speaking Greek

Aramaic Context: credible, if context suggests parties speaking Aramaic

Discourse Features: credible, if Jesus' speeches cohere in a unique style

Characteristic Jesus: credible, if it's both distinctive and characteristic of Jesus

The validity and applicability of each of these criteria can be tested with Bayes's Theorem.

DISSIMILARITY

Formulated in various ways, the Criterion of Dissimilarity has been identified as underlying many of the others, which are just particular applications of this more general principle. One reference defines it this way: "If a tradition is dissimilar to the views of Judaism and to the views of the early church, then it can confidently be ascribed to the historical Jesus."[1] Another says: "If, therefore, Jesus is presented as saying or doing things that seem almost out of place for both Palestinian Judaism and early Christianity, the likelihood of the presentation being accurate seems great."[2] Yet the latter author immediately follows this with an example that is factually false,[3] thus illustrating the most common folly in using this or any criterion: the evidence often fails to fulfill the criterion, contrary to a scholar's assertion or misapprehension. But even once you've eliminated all applications that depend on demonstrably false assertions (which represent instances of ignoring evidence or field-comprehensive background knowledge), you still have the two remaining problems I identified before: logical invalidity and inconclusive threshold.

First, we simply *don't know* much of what went on in second-temple Judaism and the early church. To the contrary, we know early Christianity and Judaism were wildly diverse and that we have scarce to no data about all the many different communities we know were flourishing at the time (I'll come back to that problem on page 129). Thus we cannot establish for alternative hypotheses either a low prior or a low consequent. Second, just because something unusual is attributed to or said about Jesus doesn't make it true. If something that unusual can happen to or be said by Jesus, then it can just as easily have happened to or been said by anyone in the Christian tradition after him. So when does "being unusual" indicate it came from an innovating Jesus, rather than a later innovating missionary or prophet? At what point does meeting this criterion make the former explanation more probable? Everything that would tend to raise the prior or consequent for a hypothesis of "historicity" on this criterion will raise the prior or consequent of the contrary hypothesis just as much—or nearly as much, producing a conclusion so far from certainty as to be of no use.

Two more particular problems arise for this criterion. First, if a purported fact was so unusual, why then was it preserved so faithfully? Any answer to the latter question will entail as much a reason to invent it as to preserve it if true. Indeed, combining this problem with the first, the most obvious answer to "why was it preserved?" is that it was entirely in accord with the views of early Judaism or the church—and we just lack the information in surviving evidence to confirm this. As Christopher Tuckett says, "The very existence of the tradition may thus militate against its being regarded as 'dissimilar' to the views of 'the early church.'"[4] The last problem is that a saying or story that *did* originate with Jesus or eyewitnesses can easily have become confused or distorted in the retelling until the version we have appears unusual against its Judeo-Christian background, yet not because it is historical, but precisely because it is not. In other words, due to our ignorance of crucial data (in consequence of the extremely poor survival of evidence), the consequent probability of any given oddity on explanations other than "historicity" can often be high— too high for "historicity" to be much more probable.

It's thus generally agreed this is one of the most deficient of all the criteria, of essentially no valid use in determining the truth.[5] And yet by underlying so many others, it takes them all down with it. I will use the following criterion (Embarrassment) as a comprehensive example of that fact, then treat the remaining criteria more briefly. Subtle variants of the Criterion of Dissimilarity (or any other criterion) are likewise no more valid, for example, the Criterion of Rarity says that if something Jesus said or did is only *rarely* evinced by Jews or Christians otherwise, then it's probably true, but that's just a non sequitur—for all the same reasons surveyed above.

EMBARRASSMENT

The EC (or Embarrassment Criterion) is based on the folk belief that if an author says something that would embarrass him, it must be true, because he wouldn't embarrass himself with a lie. An EC argument (or Argument from Embarrassment) is an attempt to apply this principle to derive the conclusion that the embarrassing statement is historically true. For example, "the criterion of embarrassment states that material that would have been embarrassing to early Christians is more likely to be historical since it is unlikely that they would have made up material that would have placed them or Jesus in a bad light,"[6] or in other words, "the point of the criterion is that the early Church would hardly have gone out of its way to create material that only embarrassed its creator or weakened its position in arguments with opponents."[7]

The EC has been much discussed, but rarely its underlying logic.[8] This is odd, since the EC is essentially an application to history of the legal principal of 'statement against interest' whose logical soundness has indeed been questioned.[9] The increasing trend in law now is to require corroborating evidence before granting admission of statements against interest. And even when admitted as evidence, juries are instructed not to assume the testified fact is true, but to critically evaluate such testimony like any other. Historians should do the same.

Adapting various formulations of the juridical 'statement against interest' rule into a concise form applicable to the EC, we can define it as "a statement made by someone having sufficient knowledge of the subject, which is so far contrary to their interests that a reasonable person in their position would not have made the statement unless believing it to be true." This sets two requirements, each of which must be reasonably proved or known for the principle to apply: we must be able to confirm the speaker was actually in a position to know the truth of the matter, and that their statement was *so far* contrary to their interests that they would not have said it unless they believed it to be true. Only in this formulation is the EC logically valid, which entails the following EC argument: "If a person made a statement so far contrary to their interests that a reasonable person in their position would not have made the statement unless believing it to be true *and* that person was in a position to reliably know what the truth really was, then their statement is probably true." So formulated, a valid EC argument is exceedingly difficult to apply in Jesus studies, where typically neither requirement can be reliably met.

In Bayesian terms, when this criterion as thus stated is met, $P(e|{\sim}h.b)$ $\ll P(e|h.b)$. So, as long as prior probability does not render the claim incredible (i.e., as long as $P({\sim}h|b)$ is not $\gg P(h|b)$), it should be believed. This diversion of the consequent probabilities is what is accomplished only when both requirements of the criterion are met simultaneously. So when one or both are not met, the argument fails to have this effect on the consequents, and thus fails to have any significant effect on the probability that the examined claim is true. The following analysis shows why we probably can never meet both requirements of this EC argument in Jesus studies.

GENERAL INADEQUACY OF THE CRITERION OF EMBARRASSMENT

When all our background knowledge about the nature of man, the world, and the ancient evidence and context is taken into account, we will find that several general defects plague the EC.

(a) Problem of self-contradiction

The first general problem facing successful application of the EC is that most actual uses of the EC in Jesus studies have been based on an inherent contradiction. The assumption is that embarrassing material "would naturally be either suppressed or softened in later stages of the Gospel tradition."[10] But all extant Gospels are already *very late stages* of the "Gospel tradition," the Gospel having already been preached for nearly an entire lifetime across three continents before any Gospel was written. The most widely held consensus in the field is that the Gospels post-date the life of Paul, as he never mentions them or any uniquely identifying information in them. And this is a strong Argument from Silence, as he cannot have been ignorant of them if they then existed, and the information in the Gospels would surely and repeatedly have been used by him or against him, either way ending up in his letters. And there is no other evidence that securely dates the Gospels to that period. Yet Paul's ministry spanned two continents and nearly three decades, long enough even for the authors of the Gospels to have been born into the Christian faith and grown into adulthood believing in no other (not that they did, but they well could have, the tradition by their time was indeed that old), while countless other missionaries (Apollos, for example) also spent decades preaching and wrangling with opponents inside and outside the church, all across Africa, as well as (like Paul) Europe and the Middle East. It's simply inconceivable that no one would ever have noticed "embarrassing" details in the story until Mark wrote them down after all this time. So the fact that we see a redactional tendency in *later* Gospels to soften or erase embarrassing material in *Mark* is very nearly conclusive proof that those embarrassing details *never existed in the tradition at all* before Mark. For had they preceded him, they would have undergone all that redactional treatment well *before* Mark put pen to paper.

We might think to make it plausible that such embarrassing material preceded Mark by proposing that it was not yet embarrassing when Mark recorded it but only came to be embarrassing to later authors. But if the material was not embarrassing to Mark, the EC does not apply, and you have no EC argument (see problem (b) below). The only remaining option is to somehow prove that Mark was under different pressures, or was innately more honest, than the other Evangelists. But the latter cannot be proved—and is doubtful, especially given the quantity of dubious material in Mark (not least being the false story of the sun going out, examined in chapter 3).[11] And the former bears consequences some scholars might fear to allow, such as conceding the other Evangelists were liars, and requiring the device of weighing down your hypothesis with conditions that explain why Mark would be compelled to tell truths that the other Evangelists felt free to suppress. As the sun-darkening story proves (among much else in Mark that's dubious, e.g., Mark 5:11–16, 6:35–52, 8:1–21, 11:13–20, etc.), the claim that eyewitnesses prevented Mark from fabricating doesn't carry much force, either. Innumerable other cases of generational storytelling establish that the commonly assumed theory of an "eyewitness check" against deviations from truth actually has little to no basis in fact. But that's a conclusion I must explore in the next volume. For the present, this one spectacular failure of the theory (Mark's inordinately public yet entirely fabricated erasure of the sun) should suffice to prove that this theory of an "eyewitness check" is in serious trouble. More to the point, such a theory could never explain Mark's failure to *omit* embarrassing information, as no "eyewitness check" would have prevented that (not even in his sources, much less in Mark).

We should also be warned against too readily assuming an author would automatically know how a statement would be taken by his audience or how embarrassing it could *become*, especially if it is later exaggerated or taken out of context. Authors often don't realize how critics will exploit the things they say, which opponents will often do in unexpected ways. It's entirely conceivable that an author could write one thing, not even thinking about how his critics will use it, then after it gets thus used, a subsequent author would be keen to try and edit the tradition to counter the new and unexpected controversy it created. The history of doctoring the manuscripts of the New Testament is rife with examples of statements that only *later* became embarrassing as they came to be exploited or emphasized in novel ways (by "heretics," which is to say, "competing interpreters of the gospel").[12] This happened even as the Gospels were composed. Like Matthew's stated excuse for introducing guards into the story of the empty tomb narrative—which reveals a rhetoric that apparently only appeared after the publication of Mark's account of an empty tomb.[13] For Mark shows no awareness of the problem. It clearly hadn't occurred to Mark when composing the empty tomb story that it would invite accusations the Christians stole the body (much less that any such accusations were already flying). Which should be evidence enough that Matthew invented that story, as otherwise surely that retort would have been a constant drum beat for decades already, powerfully motivating Mark to answer or resolve it (if his sources already hadn't, and they most likely would have). There can therefore have been no such accusation of theft by the time Mark wrote. The full weight of every probability is against it. Mark simply didn't anticipate how his enemies would respond to his story.

(b) Problem of ignorance

The second general problem facing successful application of the EC is the extraordinary degree of our ignorance. As Stanley Porter says, "determining what might have been embarrassing to the early Church" is "very difficult," especially given "the lack of detailed evidence for the thought of the early Church, apart from that found in the New Testament."[14] Even conservative scholar Mark Strauss admits that "what seems embarrassing to us may not have seemed so to the early church" and "there also may be reasons we cannot immediately recognize for the creation" of seemingly embarrassing details.[15] Theissen and Winter also call attention to this defect of ignorance: what is perceived to be in conflict with Christian assumptions and beliefs at the time of a Gospel's composition might not have been in earlier decades, and such material could have been fabricated in that earlier period.[16] Thus, perceiving material in a Gospel to be at odds with the assumptions otherwise embraced by that author cannot reliably indicate that that material is historical, any more than that it was invented by an earlier community. As Morna Hooker puts it, "Use of this criterion seems to assume that we are dealing with two known factors (Judaism and early Christianity) and one unknown—Jesus," but "it would perhaps be a fairer statement of the situation to say that we are dealing with three unknowns, and that our knowledge of the other two is quite as tenuous and indirect as our knowledge of Jesus himself." As she rightly says, "It could be that if we knew the whole truth about Judaism and the early Church, our small quantity of 'distinctive' teaching would wither away altogether," which could be equally true of our small quantity of "embarrassing" content.[17]

The incest and immorality of the gods in Homer was embarrassing to Plutarch and Plato, for example (as they chafe at it constantly), yet no one today uses that fact to argue that Homer's stories of the gods must therefore be true. Different people in different communities react to the same claims differently. And the early Christian Church was nothing if not wildly diverse, already in the time of Paul (as Paul's own letters attest), much more so by the time the Gospels were written. This same error arises when modern readers mistake what is embarrassing to *them* for what would have been embarrassing to any ancient readers, or simply get wrong what was embarrassing to ancient readers. Or they treat ancient readers monolithically, as if everyone in antiquity shared the same opinions and values. Early Christianity was in many respects rooted in open rejection of elite norms. What was embarrassing to many elites (such as embracing pacifism or placing faith before reason or even worshipping a Jewish god) was not embarrassing to Christians. That's why they were Christians.[18] One example of this fact is how rapidly Christians abandoned the elite Jewish requirement of ritual diet and circumcision, even though we know these cannot have been ideas promulgated by the historical Jesus (as in Galatians Paul reveals that he introduced that innovation himself, years after Jesus is supposed to have died; which is corroborated in Romans, where Paul makes a lengthy defense of the innovation without ever once citing the authority of Jesus).

Likewise, as Hooker observes, "what seems incoherent to us may have seemed coherent in first-century Palestine" or, I would add, in any particular Christian or Jewish community, especially if extra-textual discourse was expected to unravel any apparent contradictions or embarrassments, even turning them to particular uses.[19] The canonical Gospels together are a rhetorical nightmare of glaring problems and contradictions (indeed many of the teachings of Jesus were deliberately written to be paradoxical), yet in the end were all still embraced by a single church. In reality, most religions throughout history have buried themselves in inconsistencies, yet persist unperturbed. Just consider how much easier Unitarianism is to defend than Trinitarianism, yet how fiercely the Catholic Church clung to the latter even though it was not believed of Jesus in Paul's time. So mere contradictions (or any other rhetorical vulnerabilities) will never satisfy the EC. We need to have a firm grasp of the whole nature of a group's thinking, values, concerns, and ideas, and its particular circumstances, before we can start declaring what they would or would not have thought a useful or suitable claim about Jesus. And the fact of the matter is, we simply don't have anywhere near enough of that information for any of the diverse Christian communities before Mark. To the contrary, we know for a fact that we are ignorant of far too much.[20]

John Meier is peculiarly guilty of covertly assuming (even when overtly denying) there was no diversity or evolution in the views of the early church. Meier often speaks of the Gospels as if they were all products of a single monolithic "Church" that never changed from beginning to end, when in fact they were produced by very diverse church communities in different places and times, and thus whose authors did not share the same values and concerns, nor even the same beliefs. Hence, we cannot speak of what would have been embarrassing to "the church" (as Meier repeatedly does) because there was no such animal. There were many churches, constantly changing and competing over time and place. In consequence, what one author found embarrassing might not have been embarrassing to another, or one author may have had overriding interests that another did not, or one author may have been embarrassed by a statement to a very different degree than another, and we just don't know enough about the issues, concerns, and variations of the myriad church communities across the first century to make any confident statements about what would have been embarrassing to whom—except in a very few cases, and (as we'll see) none of those support Meier's use of the EC.

Even ascertaining what a particular author was thinking is difficult. Craig Evans frames the EC in terms of identifying a "tradition contrary to the evangelists' editorial tendency."[21] But that assumes you actually have a correct read on what that tendency is for any given author; that you actually know what that evangelist's aims were. Given the countless unresolved disagreements over this point among the entire scholarly community today, this often does not appear to be a safe assumption. And even when safely assumed, this still only moves the problem back to that author's *sources*, not all the way back to historical fact.

John Meier actually produces his own example: Jesus' cry of dereliction on the cross ("My God, my God, why have you forsaken me!?" in Mark 15:34 and Matthew 27:46), which is a quotation of Psalms 22:1. "At first glance," Meier says, "this seems a clear case of embarrassment; the unedifying groan is replaced in Luke by Christ's trustful commendation of his spirit to his Father (Luke 23:46) and in John by a cry of triumph" (John 19:30).[22] Here Meier admits that Luke and John lived in a different era and thus had "later theological agendas" different from Mark's (and Matthew's). He further admits that this cry originally served a mytho-literary function far outweighing any embarrassment it may have incurred (namely, the assimilation of Jesus in his death to a venerable Jewish tradition of "the suffering just man"), and therefore Meier concludes not only that the EC cannot rescue this statement as historical, but that the statement is probably *not* historical. "By telling the story of Jesus' passion in the words of these psalms, the narrative presented Jesus as the one who fulfilled the OT [Old Testament] pattern of the just man afflicted and put to death by evildoers, but vindicated and raised up by God," which mythic pattern Mark realized by weaving allusions and quotations of the relevant psalms "throughout the Passion Narrative," including this cry.[23]

Meier also argues that due to the mythic tradition this device evoked, Mark's play upon it would not even have been perceived as embarrassing. But here Meier's argument is immediately refuted by his own evidence that later Evangelists found it so embarrassing they changed it. But certainly his general point is still valid: that we should be careful about assuming we know what ancient readers and writers regarded as an embarrassment, particularly as this would vary even then among different communities and over time. But this is not an example of that kind of error, but of a very different one: that of naively assuming we know what an author is doing with a story. Mark's Gospel is actually rife with irony and reversals of expectation.[24] Which means Mark appears to have inserted material precisely *because* it was embarrassing—to outsiders, not privy to the "secret" (as Mark actually says in 4:11–12, 33–34). Insiders would not perceive any embarrassment, because they would be taught the real point behind everything, just as Mark says Jesus' disciples were (a story that itself establishes the model Christians were to emulate in their own churches). In effect, Mark's literature is designed to exclude people who "don't get it," thereby increasing the commitment of insiders who are made to feel they are special by the very fact that they understand what others don't. As this is explicitly stated by Paul (1 Corinthians 1:19–27, 2:14, 3:18–20, 8:1–2) and our earliest Gospel (per above), it's one of the few instances where we can be certain of how the earliest Christians thought about what they preached. We should not be "shocked" to find Mark realizing this principle, especially if you agree (as many scholars do) that Mark's Gospel is closest to Paul's in message and inspiration.

When it comes to Jesus' "cry on the cross," the fact that it begins his death with a quotation from Psalms 22:1 actually establishes a clue to the rest of the passion-resurrection narrative, which is based on a three-day sequence of Psalms (Psalm 22: crucifixion and death; Psalm 23: burial and sojourn among the dead; and Psalm 24: resurrection "on the first day of the week," Mark 16:2 quoting the Septuagint edition of Psalms 24:1).[25] Once the pattern is evident, the conclusion is inescapable: this whole narrative is a literary creation, structured with well-crafted allusions to the Psalms (so well, I believe, that we can be certain we are not reading an eyewitness record of what happened, but whether that conclusion is valid I will examine in my second volume). In Bayesian terms, the probability of all these coincidences with the Psalms is much lower on the hypothesis that they happened than on the hypothesis that Mark is creating a narrative out of the Psalms. Authors don't just randomly stumble into smartly constructed and remarkably apposite literary structures like this, nor do historical events themselves just accidentally do that for them. Thus, $P(e|\text{INVENTED})$ $>>> P(e|\text{HAPPENED})$, and since we have no evidence the priors diverge anywhere near as strongly the other way, we end up with $P(\text{INVENTED}|e)$ $>>>$ $P(\text{HAPPENED}|e)$: Mark is almost certainly inventing. Once we realize what Mark as an author is doing here, it no longer becomes plausible to see Mark as reluctantly including the "cry of dereliction" out of "embarrassment," for he could easily have invented it to suit his literary intent. By contrast, if it *were* embarrassingly true, what possible reason could Mark have had even to mention it?

Once our ignorance is dispelled, we discover the EC fails to apply. How many more applications of the EC will fail once we correctly grasp the author's actual intent? Meier fails to realize that there is no difference between this example and those he finds convincing. The evidence is identical (in every case, all Meier has to offer is the fact that later authors expose their embarrassment at the remark by changing it, exactly as they did here), and the reasoning is identical (theologically, a perfect God should not be in despair, much less talking to himself in the third person and expecting a reply, yet all Meier can propose in every other case is some similar conflict between what Christians were supposed to think of Jesus and how he was actually depicted, exactly as should be the case here). Meier thus fails to apply his own principles and background knowledge consistently. He knows full well (from his own example of Jesus' "cry of dereliction") that both this evidence and this reasoning fail to entail the conclusion intended (that the cry is historical), and yet he appeals to exactly this kind of evidence and exactly the same reasoning to argue other claims are historical, thus employing a method he already knows to be invalid.

(c) Problem of self-defeat

The third general problem facing successful application of the EC is the fact that (as I've suggested already) the EC most typically ought to entail exactly the opposite conclusion. Scholars advancing EC arguments fail to establish the answer to a key question: If a statement was embarrassing, then why is it in the text at all? You have to explain why this author included something contrary to his supposed editorial tendency. Yet any such explanation will entail as much reason to invent it as record it. The EC thus becomes self-defeating.

Quite simply, it's inherently unlikely that any Christian author would include *anything* embarrassing in his Gospel account, since he could choose to include or omit whatever he wanted (and as we can plainly see, all the Gospel authors picked and chose and altered whatever suited them—even Mark excluded a vast amount of material found in Matthew, Luke, and John, so unless that was all fabricated after Mark, Mark left out quite a lot that would have been as available to him as it was to them). In contrast, it's inherently *likely* that anything a Christian author included in his account, he did so for a deliberate reason, to accomplish something he *wanted*, since that's how all authors behave, especially those with a specific aim of persuasion. This would be all the more true for those who transmitted the tradition to Mark over the span of many decades.

It's worth remarking here, as I'll further discuss later (page 192), that everyone literate enough to compose books in antiquity was educated almost exclusively in the specific skill of persuasion: that is what all writing was believed to be *for*, and how all literate persons were taught to write. Therefore, already the prior probability that a seemingly embarrassing detail in a Christian text is in there only because it is true is *low*, whereas the prior probability that it is there for a specific, persuasive reason regardless of its truth is *high*, the exact opposite of what's assumed by an EC argument (and I can demonstrate this mathematically, see page 162). The mere fact that, as Meier observes, later Gospel authors freely omitted or revised what they received, is proof enough that that's exactly what authors do. So the fact that, for example, Mark shows no signs of being embarrassed by something that *later* authors found embarrassing, should either tell us that something has changed (which leads to the previous two problems), or that Mark had overriding reasons to include that detail. And if the latter, he (or his sources) may have had that same reason to fabricate it.

Surely if anything was actually embarrassing about Jesus, we can fairly well assume it would not survive in the record at all, since very likely no one would have recorded it, at least not faithfully. We do not have, after all, any neutral or hostile sources about Jesus from anyone actually witness to his life or (so far as we can tell) by any author in contact with anyone who was. Some of the New Testament Epistles could at least claim such contact, yet they contain nothing about the historical Jesus beyond extremely generic and essentially theological statements.[26] All the evidence that survives comes solely from Christians, hardly unbiased parties, and from several sources removed, which is not a reliable proximity to the facts. This cannot be overlooked.

For example, Craig Evans says the EC "calls attention to sayings or actions that were potentially embarrassing to the early Church and/or the evangelists," and "the assumption here is that such material would not likely be invented or, if it was, be preserved. The preservation of such material, therefore, strongly argues for its authenticity."[27] But this is a non sequitur. For if it "would not likely be preserved," then there must still be a reason to have preserved it, which reason must have been *stronger* than the impetus to discard or alter it due to its supposed embarrassment, otherwise we cannot explain its preservation at all (because otherwise "it would not likely be preserved" entails a low P(e|h.b); that is, without any reason to preserve or mention it, its preservation *is simply unlikely*). But if there were such a reason, and it was that compelling, that reason would just as easily overcome the embarrassment of a fiction as the embarrassment of a fact. In other words, that it was preserved at all entails Christians must have also had a reason to invent it that would have overcome any embarrassment it created—the very same reason they would have had to preserve it if it were true. Therefore, its preservation does not argue for its authenticity at all, much less "strongly," as Evans avers.

I'll present some examples later of exactly this problem. The exceptions are very few and hard to establish in particular cases. This means we must abandon the demonstrably false assumption that authors never write embarrassing things if they are false. Authors, especially religious ones, often have overriding reasons to include embarrassing details, such as to convey a lesson or shame their audience into action. Mark's antiheroic depiction of Peter could have exactly that intent, now modeled by modern preachers who often employ exaggerated claims of their own moral depravity and denial of God prior to "finding Christ." Such rhetoric serves as a model for redemption and as evidence of sincerity (thus simultaneously proving "you, too, can be redeemed" and "having undergone such a dramatic reversal, surely I must be telling the truth"). But sometimes we cannot even fathom the motivation. The castration of Attis and his priests was widely regarded by the ancient literary elite as disgusting and shameful, and thus was a definite cause of embarrassment for the cult, yet the claim and the practice continued unabated. No one would now argue that the god Attis must therefore have actually been castrated. The humiliation of Inanna, Queen of the Gods, was similarly embarrassing (her story even seems deliberately crafted to be), yet no one would argue from this that Inanna really was stripped naked, killed by a death spell, and nailed up in hell.[28] The mythical Romulus murdered his own brother, which was then among the most despicable of crimes, and still he remained a revered god of the Roman people—and yet no one believes that ever happened, either.[29]

We simply have no clue why these shocking stories were invented, much less became the objects of veneration and symbolic emulation. Religions frequently rally around apparently embarrassing yet entirely false myths, often in defiance of common sense. The Jews were no exception. Contrary to current assumption, the execution of their own messiah was believed to have been predicted by Daniel (Daniel 9:26; even more clearly in the Greek), yet he was widely recognized as an inspired prophet of God.[30] And the Gospels clearly regarded Daniel as an authority: Mark's apocalypse (in chapter 13) and Matthew's nativity and empty tomb stories all incorporate overt allusions to the Book of Daniel.[31] Hence it would not matter if the execution of their messiah was embarrassing, for it already had the full prior authority of God and his prophets. Which would be just as much a reason to *invent* the detail, for such an invention would overcome any and all embarrassment at the fact by virtue of having clear scriptural endorsement from God. If you could prove God had said it would happen centuries in advance, then you will get far more traction inventing a confirmation of that prophecy than you would suffer from the fact that what God had ordained was in any sense embarrassing. Whether that's what happened or not (I make no case here either way), the conclusion remains: sometimes embarrassing details are invented for a reason.

(d) Problem of bootstrapping

Some scholars admit that the EC is insufficient to establish a detail as historical, and insist it must be used in conjunction with other criteria. But as Porter observes, "this may well create a vicious circular argument, in which various criteria, each one in itself insufficient to establish the reliability or authenticity of the Jesus tradition, are used to support other criteria," with the result that "the sum of a number of inconclusive arguments is asserted to be decisive," which is a non sequitur, especially when those individual arguments are not merely inconclusive, but logically invalid.[32] Not even a million logically invalid arguments can establish a conclusion—*at all*, much less "decisively." Porter also notes that the EC is basically just a particular application of the Criterion of Double Dissimilarity, whose fatal defects Porter also catalogs, and which are conclusively brought out by

Theissen and Winter as well.[33] Indeed, some attempts to bootstrap an EC argument with other criteria can actually make that argument *weaker*, not stronger. For example, responding to instances when the EC is bolstered with the Criterion of Multiple Attestation, Mark Goodacre remarks, "I can't help thinking that one cancels out the other. If everyone, Q, an independent Thomas, Mark, Matthew, Luke all have this same material, who is embarrassed about it? The multiple attestation is itself an argument against embarrassment."[34] Quite so.

Some will say that nevertheless, in some cases the EC makes a particular claim "more probably" historical than otherwise. But even if that's true (and it cannot be assumed—you have to prove in each specific case that the EC has that effect on that data, overcoming all the problems noted so far), that by itself is an observation of no value. Ten percent is ten times more likely than 1 percent, a huge increase in probability, and yet that would still be only 10 percent likely to be historical, which means it probably isn't. In fact, the odds it isn't are then 90 percent, which makes it fairly certainly *not* historical. Thus, the mere fact of making something more probable does not make it more believable. Because it doesn't necessarily make it believable at all.

Thus historians cannot hide behind meaningless assertions like "more probably historical" in order to bootstrap their way to "probably is historical." That's logically invalid, and therefore not a rational historical argument. You have to confront the hard question: just how probable is it? And BT is the only viable method for answering that question (as I've shown in chapters 3 and 4).

SPECIFIC INADEQUACY OF THE CRITERION OF EMBARRASSMENT

Such are the general problems facing any attempt to apply the EC. Now let's examine some particular cases.

(i) Jesus' crucifixion by Romans

This is the most common example. As John Meier puts it:

> Such an embarrassing event created a major obstacle to converting Jews and Gentiles alike (see, e.g., 1 Corinthians 1:23), an obstacle that the church struggled to overcome with various theological arguments. The last thing the church would have done would have been to create a monumental scandal for which it then had to invent a whole apologetic . . . Precisely because the undeniable fact of Jesus' execution was so shocking, precisely because it seemed to make faith in this type of Messiah preposterous, the early church felt a need from the beginning to insist that Jesus' scandalous death was "according to the Scriptures," that it had been proclaimed beforehand by the OT prophets, and that individual OT texts even spelled out details of Jesus' passion.[35]

We have already seen several reasons why this reasoning is invalid. The example of the castration of Attis alone refutes it. Many religions contain scandalous claims they must work hard to defend, yet few of those claims are true. So why are we obliged to assume this one is? The fact that the OT very clearly *did* predict the execution of the messiah (Daniel is explicit, and had already convinced the Jews of Qumran) likewise refutes Meier's claim that OT support must have been sought for after the fact.[36] Maybe it was. But we cannot conclude it was from the mere fact that more OT support was accumulated over time. Certainly more and more support would be sought (especially as Christians, already convinced it's there, begin to see it everywhere). But Paul does not appear to be speaking of post-hoc rationalizations when he says the core creed of the church was "that according to the Scriptures Christ died for our sins, and that he was buried, and that according to the Scriptures he has been raised on the third day" (1 Corinthians 15:3–4), nor does Hebrews appear to be stumbling over any embarrassing historical fact when it freely assumes Jesus' death was a fully ordained Levitical sacrifice. In fact, Hebrews 9 makes such complete sense of the messiah's death, we can easily imagine that very same reasoning *inspiring* a belief in that death.[37]

Accordingly, the hypothesis that the Christian messiah's trial and execution by "the ruler who would come" (Daniel 9:26) was indeed derived from the OT is initially as good an explanation of the evidence as Meier's, particularly since the Jews at Qumran were already equating this messiah with the servant of Isaiah 52–53, wherein we have almost the whole core Gospel story.[38] This fully explains why they expected the messiah to be executed, why they would imagine (or preach) that the Romans did it, and thus why they would conceive that death as a crucifixion (it being the standard mode of Roman execution for those who have humbled themselves completely). It would only be a bonus that being hung on a stake accorded with Scripture (Galatians 3:13) and that the Jewish authorities also crucified the bodies of their convicts.[39] It's notable that later, Babylonian Jews knew only of an account in which Jesus was executed by stoning (and *then* hung up), and solely by the Jewish authorities, all exactly in accordance with Jewish law.[40] Even the Christian *Gospel of Peter* (vv. 1–2) only knows of an execution by the Jewish state, under the command and supervision of Herod Antipas, rather than Pontius Pilate. In contrast, the canonical Gospels have to twist their story into a convoluted and implausible sequence of events just to get Jesus executed by Romans and not the actual Jews who were accusing him of violating only Jewish laws. Nearly every scholar acknowledges the glaring inconsistency. Some even conclude that the story can only be a whitewash for what was really a Roman execution of Jesus for attempting a coup.[41] But one can just as easily argue the reverse, that *this* element is the whitewash, for what was really a Jewish execution for blasphemy, as the Talmud records, some Christian texts confirm, and even the canonical Gospels imply.[42]

I am not here arguing Jesus wasn't crucified by the Romans, only that we cannot establish this with an argument like Meier's. For if finding OT support *post hoc* made this "scandalous" message successful, finding that OT support *ante hoc* would be just as successful. Therefore, Meier's reasoning is unsound. We know too little about the actual thinking of Christians in the time of Paul (such as regarding the origin of their creed of a crucified messiah); there are many plausible ways a crucifixion by Romans could have been preconceived or even served Christian interests; and religions often center their message around embarrassing myths they then have to defend. So we have to go back and ask questions about the relative probabilities of competing hypotheses. The fact that Mark is deliberately casting Jesus as a "suffering just man" and packing that story with deliberate irony would be reason enough to construct the tale as it is. Indeed, the humiliation-execution theme was a trope for Jewish mythic and figurative heroes of the time, and thus *fashionable*, not embarrassing (it appears in many ancient Jewish texts, including Wisdom of Solomon 2–5, Isaiah 52–53, and 1QIsa^a 52.13–53.12).[43] If Mark thought depicting Jesus like this would be too embarrassing, he would have colored his depiction exactly as John Meier observes the later Evangelists did. That he didn't entails Mark was not embarrassed by his version of events. It served his purposes. Which means it would have served his purposes whether it was true or false. So it must be noted that contrary to common assumption, Paul never says Jesus was crucified by Romans. In Galatians 3:13, for example, he only seems aware of a Jewish execution (in which the convict would be killed and *then* crucified, the two events often conflated even in Jewish sources, like the Talmudic account of Jesus' death noted earlier). Might that be what actually happened, and Mark simply changed it up to make a literary point about Roman complicity in the corrupt world order? On what basis are we certain the answer is no?

What your analysis comes to on this detail will thus depend on a number of other facts and conclusions you must settle first. The EC argument alone is simply insufficient. The fact of the crucifixion being embarrassing would make no significant difference in the consequent probabilities, and as deifying an actually crucified man has no greater prior probability than imagining a deity crucified, there is no significant difference in priors either. Likewise, whitewashing the Jewish stoning and hanging of a blasphemer as the more heroic Roman crucifixion of a misunderstood insurrectionist has no greater prior probability than whitewashing the Roman crucifixion of an insurrectionist as the Jewish-instigated crucifixion of a wrongly accused blasphemer. Six of one, half a dozen of the other.

(ii) Jesus' birth in Nazareth

Some scholars argue that no one would invent a story about a messiah born in a hick Galilee town like Nazareth, so surely that detail must be true. Elsewhere I have refuted one underlying premise of this argument: the assumption that such an origin would be embarrassing.[44] Others have refuted its other premise by identifying several plausible reasons why Jesus would be falsely contrived as a Nazarene.

Eric Laupot makes a plausible case that the term was originally derived from Isaiah 11:1 as the name of the Christian movement (as followers of a prophesied Davidic messiah), which was retroactively made into Jesus' hometown (either allusively or in error).[45] J. S. Kennard makes just as plausible a case that it was a cultic title derived from the Nazirites ("the separated" or "the consecrated") described in Numbers 6 (and the Mishnah tractate *Nazir*).[46] As he points out, a Nazirite vow was most typically of limited duration (a fixed number of days), consecrating oneself to God by certain rituals—most prominently, abstaining from wine (which Jesus indeed vows to do: Mark 14:25; Matthew 26:29), although Kennard argues the Christians adopted the term to designate a new notion of separation or holiness reflected in the Baptist cult (similar in function to the title "Essene"). Such an interpretation is all but confirmed by Acts 24:5, where "Nazarenes" *cannot* mean inhabitants of Nazareth. René Salm notes that the Gospel of Phillip seems only to know the word as an epithet of "Truth" and not a geographical moniker (and he's right),[47] and hence it may have begun as the one and been literarily transformed into the other—quite possibly by Mark.[48] Salm's suspicion is notably confirmed by Irenaeus, who believed as early as the late second century that "Jesus Nazaria" meant "Savior of Truth" in Hebrew or Aramaic.[49] The conversion of this into a town (or its association with a town already existing) might then be just another instance of the common mythographic practice of symbolic eponymy.[50]

Since no prior author mentions a connection between Jesus and Nazareth (Paul, for example, makes no mention of it), such developments are more than merely possible. Though I do not agree with all the theories of either Salm, Kennard, or Laupot, their arguments on this point are correct: these *are* viable possibilities, at least sufficiently probable to require us to rule them out first. Judges 13:5 (in light of Numbers 6) could also have been symbolically interpreted to "derive" a fictional hometown for Jesus (in much the same way that we see wildly unexpected "readings" of Scripture in the early Christian treatise of *Hermas*). In short, Mark may have invented this detail to serve a literary purpose (as we know he did for other details in his Gospel), or he may have received a tradition that had become garbled over time, in a decades-long telephone game that lost track of the word's original meaning. That this is possible is why we must compare any EC thesis with competing explanations of how any claim came to exist. Which only BT can competently accomplish.

There is a more specific EC argument in defense of Nazareth: later Gospel writers were clearly concerned to have their messiah born in Bethlehem, where some thought the OT predicts he should have been born, but this conflicted with Mark's depiction of Jesus as a native of Nazareth, so they invented convoluted stories to explain how he could come from both places. Yet this is the same evidence even Meier agrees should be suspect: if this had originally been a problem, why didn't Mark address it? Indeed, why hadn't it already been addressed decades before the tradition even reached Mark? Why only *after* Mark do these convoluted double-origin stories arise? If Mark 1:9 is discounted as an interpolation (via contamination from, or harmonization with, the other Gospels, which we know to have been a frequent occurrence in their transmission), then Mark never actually said Jesus came from Nazareth. In fact Mark seems to imagine him hailing from Capernaum (cf. Mark 2:1 and 9:33 with 6:3–4), which was also in accord with prophecy (Isaiah 8:21–9:2, verified by Matthew 4:12–16).[51] The epithet 'Nazarene' might then not be geographical even in Mark, but instead mistaken for such only by *later* authors.[52] But either way, we still have no credible case for an EC argument using only Mark. And since, as far as we can tell, later Gospels only get the idea that Jesus came from Nazareth from Mark, we have no EC argument at all.

So although Matthew and Luke's struggle to make Jesus come from both Nazareth (as Mark claimed or implied) and Bethlehem (as the OT predicted) suggests a Nazareth origin was embarrassing to Matthew and

Luke, there is no indication it was embarrassing in Mark. The embarrassment seems to have been created by Mark's introduction of a Nazareth
origin, combined with a need to have Jesus fulfill a prophecy of origin. It
seems only later Christians may have evinced that need. Yet Matthew even
claims a *Nazareth* origin derived from prophecy—in fact, not the town,
but the epithet, which could thus have been Mark's source as well (see
Matthew 2:23).[53] Unless Matthew was lying, we are obliged to agree that
Mark (or his source) could have had that same prophecy in mind. Thus we
have two different prophecies, with later Christians trying to force Jesus
to fit both, but they have exactly the same reasons to invent either. So if
Matthew and Luke are inventing a Bethlehem origin to force a fit with
prophecy (as this argument for the authenticity of the Nazareth tradition
entails), and Nazareth was also prophesied (possibly in a now-lost scripture, like the *Hazon Gabriel*, or a lost variant reading in an extant OT text),
then Mark had as much reason to invent a Nazareth origin to force a fit with
prophecy as Luke and Matthew (or their source) had to invent a Bethlehem
origin.[54] Even if this began as an epithet later interpreted as a town the
same process could occur. There is an apposite parallel in Matthew's dupli-
cation of the donkeys Jesus rode into Jerusalem: trying to force the story
to fit what Matthew took to be a double prediction, he invented an extra
donkey (implausibly having Jesus straddle both), but this in no way argues
that the first donkey must then have been historical.[55] To the contrary, it
was already itself mandated by prophetic Scripture. Thus the fact that
Matthew found just the one donkey "embarrassing" does not argue for that
one donkey being historical.

So here the EC argument collapses under the weight of the problem of
ignorance. It also falls to the problems of self-contradiction and self-defeat,
since the argument requires us to imagine an embarrassment that persisted
for decades was only resolved *after* Mark, as if Mark knew nothing of
it, which contradicts the assumption that it was ever embarrassing before
Mark wrote of it. Why, after all, does Mark even go out of his way to use
the words *Nazareth* or *Nazarene* in the first place? His story appears to
operate quite well without them. So if they were embarrassing, he could
have simply omitted them. That he didn't entails he was including them for

some specific purpose, and whichever purpose that was, it would serve a fiction as well as a fact.

Here our background knowledge establishes that weird corruptions and attributions like this occur frequently in mythic and legendary traditions (hence many a mythic hero was given an obscure town to hail from), so prior probability doesn't favor either hypothesis (unless we can adduce more evidence otherwise—hence the EC cannot stand alone), and there are several plausible hypotheses *other* than "historicity" that render the Nazareth detail every bit as likely (if not more so, for example by better explaining the absence of this detail in any earlier Christian literature, as well as the unusual use of it in Acts and the Gospel of Phillip, and the testimony of Irenaeus), leaving no relevant difference in the consequents. Thus the EC cannot establish that Jesus was born at Nazareth. Again, perhaps some other argument can, but the EC is unable to.

(iii) John's baptism of Jesus

We've dealt with supposed embarrassments regarding Jesus' birth and death. The next most common example falls in between: his baptism, ironically a symbol of both death and (re)birth. As John Meier puts it, "the baptism of the supposedly superior and sinless Jesus by his supposed inferior, John the Baptist," who was proclaiming "a baptism of repentance for the forgiveness of sins," must have been embarrassing, because it contradicts Christian beliefs (that Jesus was "superior" and "sinless"), and because subsequent evangelists scrambled for damage control.[56]

But this is the same double error Meier himself refuted in the case of Jesus' cry on the cross. First, we might see subsequent evangelists were embarrassed by the story, but Mark is not—had he been, he would already have engaged the same damage control they did. In fact, this would have been done by transmitters of the story decades before it even got to Mark (probably even before Jesus had died). The embarrassment is thus obviously new. So on that point alone the EC fails to apply. Second, Meier simply assumes Mark (and all prior Christians) believed Jesus was "superior" and "sinless" and thus would not countenance anything implying otherwise.

Neither is even plausible, much less established for early Christians, or even those of Mark's time. Paul included Christ's voluntary submission and humbling as fundamental to the Gospel (Philippians 2:5–11). Christians did not imagine Jesus as then "superior" until he was exalted by God—at his resurrection (e.g., Romans 1:4, 1 Corinthians 15:20–28). There is nothing in Mark's depiction of Jesus submitting to John that conflicts with this view. Only *later* Christians had a problem with it. Mark instead portrays what Christians originally thought: that Jesus would be exalted as the superior *later on*. Hence he has John say exactly this (Mark 1:7–8). Likewise, the notion that Jesus was "sinless" from birth is nowhere to be found in Mark or Paul. It is clearly a later development, and thus not a concern of Mark's. To the contrary, Mark has full reason to *invent* Jesus' baptism by John specifically to *create* his sinless state, so Jesus can be adopted by God, and then live sinlessly unto death. Mark makes a point of saying John's baptism remits all sins (1:4), that Jesus submits to that baptism, and that God adopts Jesus immediately afterward (1:9–11). This is hardly a coincidence. The role of John and his baptism are explicitly stated by Mark: to prepare the way for the Lord (1:2–3). And that's just what he does. There is no embarrassment here. Again the EC does not apply.

So when Meier insists "it is highly unlikely that the Church went out of its way to create the cause of its own embarrassment," we can see in fact such a thing is not unlikely at all: once Christians started amplifying the divinity and sinlessness of Jesus, the story Mark had already popularized started to create a problem for them, so they had to redact that story to suit their changing theology. This proves Mark preceded those redactors and lacked their concerns, but it doesn't prove Mark's story is true. To the contrary, Mark had a clear motive to invent the story, particularly as he needed to cast someone as the predicted Elijah who would precede the messiah and "reconcile father and son" (Mark 1:6 in light of 2 Kings 1:8; and Mark 9:11–14 in light of Malachi 4:5–6) and set up Jesus' cleansing for adoption. Why not cast in that role his most revered predecessor, John the Baptist? Having John prepare Jesus by cleansing him of sin and establishing his divine parentage, and then endorsing Jesus as his successor, is actually far too convenient for Mark. That is *not* a statement against interest![57]

As John Gager explains, "in Mark, the incident appears very briefly and with no sign of embarrassment or editorial 'improvement'" or "discomfiture" of any sort, and therefore the EC doesn't apply, especially as there are plausible reasons for Christians to have invented the detail.[58] Gager cites Enslin's theory that the narrative allows Christians to co-opt the authority of the Baptist cult by representing their leader as his designated heir. Another reason for inventing the story is more symbolic (yet perhaps more obvious): this baptism, as for all other Christians, represents for Jesus "rebirth" through adoption by God. Just as Christ's crucifixion and last supper are both models for Christian life, so is his baptism. That it was intended as such cannot be doubted, given that it had been the Christian tradition for decades already that that's what a baptism was: being reborn as an adopted son of God (Romans 6:3–4; 1 Corinthians 12:13; Galatians 3:27–29, 4:5–7). Mark could hardly have told a story about Jesus' baptism and meant anything else by it (at least without explaining to his readers that he did). This is corroborated by Mark's ending the baptism with God's declaration of fatherhood, as well as the implications of his quoting Psalm 2:7, which evokes rebirth.[59] If Jesus had to be reborn through baptism as a model for all other Christians, whom could Mark have chosen to baptize him? Was there any more suitable choice than the famous John the Baptist?

With three solid theories of why and how Mark would invent this baptismal tale, each quite probable, the EC fails. It's not even clear that Matthew and Luke were actually embarrassed by Mark's story. The notion that Luke tries to downplay John's baptism of Jesus is only based on subjective assumptions about Luke's ordering of verses. But if we set aside our preconceived suspicions and just read the text, Luke concedes the baptism without blush (Luke 3:7, 12, 21). And though Matthew inserts an apologetic vignette (Matthew 3:13–15), this really only clarifies what Mark already said (Mark 1:2–3, 7–8). Though Matthew's insertion does bear signs of a post-Markan magnification of Jesus, he is still just responding to a criticism that Mark had not foreseen would result from his story. That Matthew thought he'd eliminated the embarrassment by making clear what Mark had already meant suggests the baptism itself was not embarrassing, even to Matthew—merely its interpretation could be. This being an instance of unforeseen embarrassment, once again the EC fails to apply. Only the Gospel of John takes the logical step of actually deleting the baptism (while retaining most of the other connections Mark had established between Jesus and John: e.g., John 1:29–34). So perhaps we can say the authors of *John* found the baptism embarrassing, but now we are so far removed in time from Mark we have no valid EC argument left to make. For John's theology could not be further removed from Mark's—indeed, in John, Jesus is identical to God (John 1:1–5) not (as had originally been preached) his subordinate (Philippians 2:5–11; 1 Corinthians 15:24), eliminating any rationale for the baptism. Again, even at best, prior probability would favor neither the historicity nor invention of this detail, and no difference remains in the consequents; while at worst, we've already seen Mark is not developing a reputation for straight-up, reportorial honesty, tipping the priors toward fabrication, while the wholly suspicious convenience of both the claim and the structure of its narration in Mark, and the absence of this claim in any prior author, tips the consequents well in favor of fabrication. So the historicity of the baptism cannot be established with the EC. To the contrary, in the final analysis it looks quite dubious.

(iv) Jesus' ignorance of the future

I have shown how the three EC arguments most often regarded as unassailable are in fact unsustainable. Any other example you care to choose will fall to the same analysis. One such is Mark 13:32, where we find, as John Meier puts it, "the affirmation by Jesus that, despite the Gospels' claim that he is the Son who can predict the events at the end of time, including his own coming on the clouds of heaven, he does not know the exact day or hour of the end." But Meier gives no explanation why we are supposed to believe Mark expected Jesus to be omniscient.[60] Mark simply says Jesus knows some details and not others; that God has reserved those for himself and not told his Son. Mark shows no embarrassment at this at all. That would only be embarrassing to later Christians who were increasingly equating Jesus with God, to the point that it became less and less intelligible how God could simultaneously know and not know something. Hence, as Meier shows, later scribes tried meddling with this passage in Mark and Matthew, while Luke and John deleted it altogether. But this concern did not exist in Mark's time, or before (as surely it would have been deleted decades before Mark even got ahold of the story). According to Paul (in the verses cited on page 148), Jesus was an appointed emissary of God, not identical with him, and there is no evidence Mark thought otherwise (in fact quite the contrary: Mark 10:18, 14:36, etc.).[61]

This is another instance of Meier simply ignoring the fact that the embarrassment was only created by a development in Christian theology that occurred long *after* the embarrassing statement had been made. Since it was not embarrassing when made, the EC does not apply, and therefore there is no EC argument for the historicity of Jesus' ignorance of the exact time of the apocalypse. To the contrary, when this statement is examined with the correct logic, we must consider why Mark would even bother including this remark from Jesus. Why not simply omit it if it's supposed to have been embarrassing? That's what Luke and John did; so why couldn't Mark? A moment's thought should lead us to a far more obvious explanation. Mark is *inventing* this ignorance as an apologetic to explain what *was* embarrassing to Mark and his community: the fact that the end had not yet arrived, even though everyone up to then had believed Jesus told them it was nigh. By having Jesus declare the hour was not yet known to him, Mark was rescuing Jesus from the most fatal charge of being a false prophet (or at least rescuing the received tradition about Jesus from this charge, since it is not simply a given that the "end is nigh" is what Jesus actually taught, rather than coming from Christian prophets who, like the author of Revelation, claimed to be communicating with his resurrected spirit).[62] It's ironic to see what Mark probably invented to *counter* an embarrassing fact now being used as if it were itself so embarrassing that what Mark invented had to be what Jesus actually said!

In this case, the EC not only fails, but the conclusion should be quite the reverse. There is no reason to favor either hypothesis with a higher prior (at least on the facts given here), but the evidence as a whole is more probable if the statement is fabricated than if it's historical (there being no likelihood of its inclusion otherwise), so the balance of consequents ultimately favors fabrication.

(v) Did Jesus not know he was the Son of Man?

In Mark, Jesus speaks of an eschatological Son of Man as if it were a
different person. Some argue this proves Jesus really preached that,
because later Christians (as we see in subsequent Gospels) believed Jesus
and this eschatological Son of Man were one and the same, and thus would
never depict Jesus saying the contrary.[63] But Mark alone uses the concept
of a *mysteriously unidentified* Son of Man, a mode of speaking that suits
Mark's narrative theme of a "messianic secret." Hence it's just as likely
(if not more so) that Mark has Jesus speak of the coming Son of Man as
if he were speaking of another, when secretly he was speaking of himself.
Since Paul already tells us that Jesus and the coming eschatological Lord
were one and the same decades before Mark wrote (Paul speaks of no
Son of Man or anyone else returning but Jesus: 1 Corinthians 15:20–28;
1 Thessalonians 4:16–17), it is not believable that Mark would think
otherwise, much less say otherwise. And in fact Mark clearly understood
them to be the same person (Mark 8:31, 9:9, 9:31, 10:33, 10:45, 14:21,
14:41). If he intended the contrary he would certainly have to be keenly
aware of the embarrassment (or indeed confusion) it would cause, as
would those transmitting the material to him—so why did neither he nor
they seek to resolve the problem, or even acknowledge it? Again, the EC
becomes self-defeating: it's so improbable that Mark would mean they
were different people (when Paul knew no such thing, and Mark clearly
meant them to be the same) that we can be sure he didn't. We should
therefore conclude Mark is using a literary device, not recording the true
words of Jesus (unless Jesus really did preach in this mode of third-person
secrecy, though even then he would still be speaking about himself).
Of course, it's also possible Mark perceived no contradiction in using a
Son of Man source text that didn't even come from Jesus (but derived
instead from some other Christian or even pre-Christian prophet) and then
attributing it to Jesus, but that would still entail he felt no embarrassment
at the result, and in any case the sayings would then *not* be historical. So
no EC argument can gain purchase from that premise, either.

Again, the EC fails here, and only BT gives us a sound argument. If this detail were at all embarrassing, prior probability would favor fabrication, as historicity would make suppression or alteration far more probable than faithful retention (for a mathematical demonstration, see page 162). And if this detail *wasn't* embarrassing, the priors could go either way, but the consequents would then slightly favor fabrication: because it's more likely Jesus would be depicted saying this in Mark if Mark were employing a literary device well known to him (or Mark's source had done so), than if Jesus really said such things, particularly as the latter does not make probable any of the other evidence (such as in Paul), while the former does. So a valid and sound analysis argues *against* the historicity of these sayings more than in favor of them (even if not decisively).

(vi) Jesus' betrayal by Judas Iscariot

In the hypothetical source document Q, Jesus is made to say all twelve of his disciples would receive eternal honors (Luke 22:30; Matthew 19:28). Meier insists:

> If one wants to claim that the saying was instead created by the early church, one must face a difficult question: Why would the early church have created a saying (attributed to the earthly Jesus during his public ministry) that in effect promised a heavenly throne and power at the last judgment to the traitor Judas Iscariot?[64]

Even Theissen and Winter declare, "Early Christianity always numbered Judas Iscariot among the twelve disciples and had simply scorned and condemned him as the one who betrayed Jesus," so "When nevertheless the Jesus tradition preserves a promise that the twelve (and not the 'eleven'!) disciples will exercise future rule over the restored Israel (Matthew 19:28/ Luke 22:28–30), there can be no doubt that this is a saying that has withstood the tendencies of the tradition."[65]

Already, their first fact isn't true. For at least the twenty or thirty years of Paul's ministry, in other words the entirety of "early Christianity," we never hear of *any* of these claims (that a Judas Iscariot was one of the twelve or that he, or *any* member of the twelve, betrayed Jesus or was "scorned and condemned" for doing so, despite an occasion to mention it in 1 Corinthians 11:23–27; nor even the saying about twelve thrones, despite an occasion to mention it in 1 Corinthians 6:2–4). If anything, we have evidence confuting this betrayal. Unless we admit to an interpolation, Paul says "the twelve" were honored with a vision of Jesus almost immediately after his death (1 Corinthians 15:5), which is hard to reconcile with any notion that one of those twelve was known to have betrayed Jesus. This likewise suggests the later Gospel story of Judas's suicide must be false (since he must then have been still alive to receive—and report—a revelation of the risen Jesus). It's therefore more likely that the story of Judas's betrayal is a literary invention, whose meaning was thought more important than any embarrassment it might cause.

What Meier, Theissen, and Winter might then wish to argue is that Jesus' statement that "the twelve" would reign in the future world contradicts the *literary invention* of Judas's betrayal. But those two traditions may have separate origins: the prediction does not exist in Mark, who (so far as we know) invented the Judas story, while the Judas story does not exist in Q, the hypothetical sayings source that alone included any reference to a rule of the twelve—unless Matthew 27:3–10 and Acts 1:18–20, the tales of Judas's suicide, derive from Q (despite being so oddly contradictory), but even if Judas's betrayal found its way into Q, it's still widely believed Q was redacted several times (and the Judas story certainly wasn't part of any collection of sayings from Jesus), leaving still the question of whether Judas's suicide was a later addition to Q or derived from yet another source.[66] Likewise, the predicted rule of twelve may have originally been communicated in visions of the risen Jesus (and thus might not derive from a historical Jesus) or even been adopted from pre-Jesus messianic tradition, all before the Judas story was conceived. The fact that Jesus does not say "the twelve" will rule but that those who follow him will sit on twelve thrones judging the twelve tribes seems innately disconnected from the number of disciples (since the governing number is the Jewish tradition of twelve tribes, not the number of people who just happened to be following Jesus). Hence it seems rather to derive from typical messianic thinking of a sort that would not require Jesus to have ever said this himself. It would also have been in the interests of "the twelve" (if such a body really was known to Paul) to have invented this saying for Jesus, specifically to legitimize their special status (or even to have taken it from a pre-Christian apocalypse and attributed it to Jesus for the same reason). On the other hand, if there was no "twelve," the Q statement no longer has any connection with the disciples at all—it then is simply another apocalyptic prediction (that twelve men someday elected by the messiah would sit on twelve thrones), which could have derived from any source and been later attributed to Jesus (like so many other sayings may have been).

In any of these scenarios, the myth of Judas's betrayal would have arisen in a tradition separate from the "twelve thrones" saying, and these traditions were combined later—creating a new problem that (like so many others) wasn't noticed right away. And yet it's just as possible the saying was simply invented by Matthew, without realizing the contradiction it thus created—or without believing it *did* create a contradiction. For if Matthew believed that, despite the fate of Judas, there would nevertheless be an official "twelve" to enthrone when the time came, as both Luke and Paul seem to have believed, then no contradiction results.[67] After all, Matthew is smart enough to emend the number to eleven when describing the first resurrection appearance to the disciples (Matthew 28:16; likewise Luke 24:9 and 33, and Pseudo-Mark 16:14). He would not suddenly forget about the same conflict remaining with the twelve thrones saying. The odds that at least one of all these scenarios occurred are not small enough to grant historicity a significantly higher probability. They all have nearly the same prior probability, and none far enough outstrips the others in consequent probability.

When we return to the supposed "embarrassment" of inventing the story that a member of the twelve betrayed Jesus, we find that argument equally weak. As noted already, the evidence makes it unlikely that any member of the twelve, much less named Judas, engaged in any such betrayal as Mark depicts. No one seems to have heard of this before Mark. The betrayal story also makes no historical sense. The authorities did not need Judas (much less have to pay him) to find or identify Jesus (Mark 14:10–11, 14:43–50). Given what Mark has Jesus say in 14:49 (and what

Jesus had been doing in Jerusalem only days before), the authorities knew what he looked like, and they could have seized him *any* time he appeared in public. They were not on a timeline. The idea that Jesus had to be tried and crucified illegally in a rushed overnight trial exactly at Passover is a Christian theological concept that cannot have had any role in the decisions of the Sanhedrin (especially since they had jails to hold him over in). Thus, the story as a whole looks like fiction.[68] The inclusion of Jesus' foreknowledge of Judas's betrayal (Mark 14:17–21), directly in parallel to his prediction of Peter's betrayal (14:29–31), both framing the Eucharist and prediction of resurrection and apostolic abandonment, only highlights the mythic character of the entire plot element. Therefore, that it looks (or later became) embarrassing is not necessarily because the story is true. Indeed that must be improbable, as the subsequent trend was to make Judas's character and betrayal even *more* despicable (and hence more mythically grandiose), rather than apologetically softening or eliding it or explaining it away (or even, in fact, making it any more historically intelligible, which confirms later redactors had no genuine sources). This suggests Mark's invention of Judas's betrayal would not in fact have been embarrassing, because it was something later authors found rhetorically *useful*, and even amplified. And if them, so Mark.

The fact that Jesus' betrayer's name essentially means "Jew" should already make us suspicious.[69] Mark may have intended him as a symbol of particular recent poignance. In both name and deed, Judas may be an intentional symbol of the very internecine betrayal that was destroying Jewish society and causing it to fail to realize God's kingdom, even just recently having caused the destruction of Judea, Jerusalem, and God's own Temple (if Mark wrote in the 70s CE, as most scholars now think). Judas was also a name famously associated with the path of violent rebellion (Judas Maccabeus and Judas the Galilean), which is all the more obvious an allusion if "Iscariot" is (as many scholars believe) an Aramaicism for the Latin "Sicarius," the infamous "Killers" whom Josephus blames for provoking Rome to bring about the destruction of the Jews (which would further mean that Judas's full name meant in Aramaic "The Jew Who Kills [Him]," which one might think would be too coincidental to be histor-

ical). The name Judas may also be intended to evoke the divided king-
doms Judah and Israel, a symbol of Jews disunited and at war with each
other, the more so if you agree that a number of indicators suggest Jesus
is typecast in the Gospels as a symbol of Israel (as Thompson argues in
convincing detail in *The Messiah Myth*), which alone could have inspired
the creation of a Judah to oppose him. The text of Zechariah from which
Matthew borrows many details of his expanded Judas story even contains
this very juxtaposition, including the very name of Judas.[70] In Zechariah,
the one who is paid the thirty shekels is to "become shepherd of the flock
doomed to slaughter" (an apt description of Judas in respect to Jesus) and
then, by abandoning the task (and the sheep to their death) and casting the
money aside, to "break the brotherhood between Judas and Israel" (the
very point of the Judas story: you can take the money and die, or follow
Jesus and live, thus either joining the New Israel or the grave).[71] Matthew
thus saw this very symbolic value of the Judas story, which inspired him
to exaggerate it with even more scripturally derived detail. Mark may have
had the same idea all along. So whether this possibility is at all probable
must first be explored before we can rule it out (I'll examine it again in my
next volume).[72]

So even here an EC argument gets little traction. Instead, the consequents fall in favor of fabrication for the Judas story, and even slightly in
favor of fabrication for the twelve thrones saying, since the hypothesis of
"fabrication" makes all the evidence more probable. If the priors favor
neither, the consequents prevail, and we should conclude the Judas story
is myth, and the twelve thrones saying only *possibly* authentic at best. Any
attempt to gainsay this conclusion requires presenting evidence that ups
the prior or consequent for historicity, or lowers either for fabrication, thus
even if you disagree with this conclusion, only BT can lead you to the
correct one. The EC alone is of no use.

And so on . . .

Similar arguments eliminate every other attempt to deploy an EC argument
on the Gospel materials. That Jesus had enemies who slandered him, that

Jesus went to parties with sinners to save them, that Jesus' family rejected him, and so on, all face the same problems of self-contradiction (had they been a problem, they would have been removed or altered long before Mark even wrote), ignorance (we don't really know whether these stories were embarrassing to the communities who told them at the time they were first told), and self-defeat (any reason to preserve them if true can be just as much reason to fabricate them, and in every case we can easily construct plausible motives for their invention, which often make even more sense than the stories being true).[73]

When EC arguments are scrutinized, some fall to the analysis that the experiences of Christians themselves in their battles, trials, and evangelizations were being mapped onto Jesus as a model to follow and commiserate with. Indeed, to imagine God suffered the same things you do is the highest form of vanity, hardly a statement against interest. Such constructs, moreover, are always rhetorically useful, and thus always well motivated. For example, Jesus being called crazy aligns too well with the fact that Christians themselves faced this charge—so how apposite to depict their Lord as being unjustly accused of the same, and then supplying him with clever speeches refuting it. That's simply too useful to be a statement against interest.[74] Christians similarly faced conflict from their families,[75] which statistically must have involved on occasion the same charge of insanity or demonic possession from them; so depicting their Lord as trading his family in for a new one in result (Mark 3:21–35, and that in the very same scene), is again too convenient. Similar tactics have been employed by many a cult throughout history, dividing members from their established family, and representing the cult as their new "true" family. So "Jesus did it, too" would not be a statement against interest. To the contrary, it reinforces exactly what Christians wanted to preach (hence Luke 14:26).

Other EC arguments fall to the analysis that the evidence already argues the premise is false, whether the EC applies or not, like Jesus' betrayal by a member of the twelve. Still others fall to the analysis that an embarrassment cannot even be established. The claim, for example, that Mark would not invent a story about women being the first to discover the empty tomb because the testimony of women would be too embarrassing, is based on claims about the ancient world that are simply not true.[76] Other EC arguments fall to the analysis that the facts depicted are so unbelievable that no matter how embarrassing they may have been we can still be certain they were never true. As Dennis MacDonald observes of the unbelievable fickleness and stupidity of Jesus' disciples, an EC argument cannot sustain the belief that they were really like that, because no human beings are. The depiction is so contrary to any plausible reality it can *only* be fiction. And MacDonald makes a good case for what the literary function of that fiction was.[77] Paul Danove confirms MacDonald's point with his own demonstration that this is yet another example of Mark's deliberate use of irony to make a point.[78] Here, it's much more improbable that such stories would be true than that any embarrassment they caused would prevent them being told. In other words, the consequents favor invention, not historicity.

There are also some EC arguments that fall to the analysis that even though an event may be realistic in principle, its depiction is so literarily crafted we should be far more suspicious than the EC would have us be. Peter's thrice denial, for example, reads like a morality play, as if something out of Shakespeare rather than real life, with dramatic features more befitting a novel than a history. That his three denials (a fact odd to focus on in such detail) contrast too appositely with the women's three acts of loyalty (attending the cross, burial, and resurrection) is likewise suspicious. We have to ask why this story is told at such length and with such unusually meticulous detail—indeed, why it is told at all. Mark would not have included this story (much less composed it so carefully) unless he wanted to. So why did he want to? The answer is unlikely to support its historicity. Even assuming the naive view that, as tradition claims, Mark is simply recording what Peter preached, why would *Peter* be preaching this? Could it be because he found the story useful as a missionary? Yet such utility would attach to a fiction as much as a truth.

That all these EC arguments fail does not entail these claims are unhistorical. They may yet be rescued by other arguments. But that can only be accomplished by applying BT.

WHAT ARE WE TO DO?

So when is an EC argument successful? Certainly it isn't always unsound or invalid. It can work in a court of law, and often finds successful use in every historical field (even if called by other names). I've relied on it myself. But the only way to deploy a successful EC argument is to avoid or overcome all the problems surveyed above and then produce a logically valid and sound argument.

First, you must reliably know if the statement in question very probably did go against its author's interests, that the author actually perceived that it would, and that the statement did not serve other interests the author had which he may have regarded as outweighing any other consequences he perceived to be likely. And that means you must reliably know what an author's interests actually were, and not just in general, but *that* particular author, in *that* particular book, in *that* particular scene (and in that particular community at that particular time), and you must reliably know what the author perceived the consequences of his statement would be—not just what they actually were, because an author might have not foreseen those consequences, or underestimated their severity. And then you cannot simply rest on the expected negative consequences of a statement, because an author may have had overriding interests. So you must reliably know how that author would have weighed the pros and cons he was aware of at the time, especially if he thought the cons could easily be explained away or overcome in extra-textual discourse. And if you can establish all that, you're not done. For you must also reliably know if the author was even in a position to know the statement was actually true. Because that an author believed it was true does not entail it was.

Meeting all these conditions can be difficult, especially in the study of Jesus. You also need a specific theory as to why the questionable statement was included *at all*. And then you need to test that theory against other theories of why it may have been included—at the very least, you must test it against the most likely of those alternatives. You have to ask: even if it was historically false, what is the most likely reason that that statement could have been included, by that particular author, in that particular book, in that particular context? The answer to that question is the theory you will test against your own, which you must also spell out exactly—this being usually that the most likely reason the statement was included was that it was true but this author couldn't omit it or change it despite having ample reason to on account of its embarrassing nature. But that requires explaining why that author *could not omit it or even change it* (and why no one else could, in all the decades before). Because some explanations of that odd fact will be far more far-fetched than others (and for any explanation, the more far-fetched, the lower its prior probability must be), and because you are obligated to prove that that explanation was in fact guiding that author's construction of the text (because, again, the less certain you are that it was, the lower its prior probability must be), and because some explanations will provide just as much reason to invent a fact as to have reported it if true (which then renders your theory's consequent probability no higher than the alternative).

In other words, you can't just look at some statement *x*, check the box "EC applies," and then conclude "*x* is probably historical." You have to show that the EC not only actually does apply, but also that it actually does have this effect. And to prove the latter you first have to show (or be able to show) that the EC increases the probability of *x* at all, and *then* you have to show (or be able to show) that it not only increases it, but increases it enough to bring it to some degree of historical certainty. And you must do this in a logically correct way—which requires taking alternative explanations into account, and all available evidence and background knowledge. And even after achieving all of that, you cannot fallaciously tout the false dichotomy "true or false," but you must honestly acknowledge degrees of certitude, that is, is *x* only somewhat probable, or very probable, or nearly certain? Because it makes a substantial difference which is the case, particularly if you plan to use *x* as a premise in another argument. Historians cannot hide from all these obligations. Because they all have consequences, and you cannot responsibly hide from the consequences of your own arguments and assumptions.

Hence to get an EC argument to work, we must first answer the question of when *any* method of criteria is logically valid and sound. Any method that conforms to BT will be valid. So as long as its premises are then all soundly demonstrated, a BT-structured EC argument will be valid *and* sound. Applying the logic of BT to the EC, we must first ascertain if our author is just reporting the facts as witnessed by or told to him (or what he purports to be those facts) or constructing a story for some purpose *other* than making a record of what happened (such as producing myth, fiction, or any other form of storytelling whose aims are more subtle than superficial veracity). For if the latter is the case, the EC no longer applies at all, because then the probability that an author will knowingly include *any* embarrassing statement is practically zero, as he will only include what he *wants* to include. Which means if something got included, it cannot have been a statement against interest. Only in the former case (of attempted historical reporting) will someone include anything like statements against interest, either out of sincere neutrality or in the effort to make sense of unexpected facts or because it cannot be avoided—or some equally compelling reason.

But even in a case of genuine reportage, authors still limit what they say to what serves their interests. So it is still unlikely they will include any embarrassing statement unless they need to or want to for some reason. And again, such needs and desires might just as easily motivate a fabrication. And if not that, then such needs and desires will still entail the author will most probably have protected his interests by apologizing for the statement, or defending it, gainsaying it, refuting it, or attempting to spin it, thus entailing predictions about the way the evidence will be presented, which, if that's not what we find, entails a lower consequent probability. So we must attend very carefully to the context in order to ascertain *why* the author included a seemingly embarrassing statement, in order to ascertain if he really was forced to against his interests, or if that statement instead *served* his interests, or if we should expect the evidence to have been presented differently (if it didn't serve his interests to report it without comment). And remember we cannot merely "declare" the author had some particular reason to report it—we are obligated to *demonstrate* that that reason was in fact operating on that author, or probably was. Otherwise, it's no more likely than any alternative motive.

Attending to all that we must address the three premises of BT. The first is prior probability. For that you must answer both principle questions: whether the author was writing myth or, if writing history, doing so with remarkable candor. In other words, you must ascertain how frequently that author, in that document, particularly in analogous literary and narrative contexts, fabricates data rather than reports what he learned from reliable historical sources. If there are many instances of doubtless fabrication or the use of unreliable sources or methods, then the prior probability that this author reports data because it's true (rather than only because it suits his story, or his source's story) is low *even if he purports to be recording what happened.* Thus, even if you can establish he was writing history and not myth, ascertaining that intent can only get you halfway. You must still ascertain the degree of his honesty (and the degree to which he verified and thus really knew the truth—rather than merely repeated what he was told). And though many scholars wish to avoid the question, the fact of the matter is the Gospels provide considerable evidence that their authors' honesty was not exemplary (and there is no clear evidence of their taking pains to verify anything, either, or even of their having reliable sources at all).[79] And if you grant that, then you must further grant that it is very unlikely anything that seems embarrassing was included because it was true, for such an author (or his source) more likely had a reason to include it even if it was false—otherwise (once we've granted their lack of candor) we must confess they would not have included it at all (and if an author shows no effort at checking facts in any reliable way, even if he's being completely honest he may be repeating a fiction, by merely *believing* it a fact, or having no way to confirm it's not).

If, on the other hand, you wish to deny this conclusion, and insist instead that the Gospel authors spoke with remarkable honesty (and reliably researched every detail), you must prove this first—a daunting task, given that we can demonstrate that they knew of each other's work, and also cannot have been ignorant of the same traditions the others recorded (if so they did), and yet still contradict each other freely, without even acknowledging the contrary accounts. Otherwise, how did some authors know the traditions they record, while other authors never heard of them?

To claim that one author had access to a tradition not passed down to the social circles of another author is to admit the Christians were working with isolated, unchecked, and highly deviant traditions, which could *not* be corrected if false as other Christians didn't even know of them. Which then pushes back the question of an author's honesty to that of their sources, which is even more difficult to evaluate, because these authors never tell us who their sources were, or how they corroborated what they said (or even whether they did)—hence we have no information at all as to the honesty and reliability (or even interests) of those sources.

So unless you have demonstrated otherwise (and I would argue no one sufficiently has in this case), the prior probability that what seems to be an embarrassing statement in the Gospels is true is actually *low*, not high. In any friendly tradition, most authors simply will not have included embarrassing truths; but all will have readily included embarrassing myths that served a literary purpose, since the motive is reversed: authors *want* to include such stories, not to suppress them, unlike embarrassing truths, most of which authors will want to suppress. The frequency of embarrassing myths in friendly sources that serve a literary purpose is essentially 100 percent. Because such myths will *only* be created and repeated to serve a literary purpose, therefore all of them will serve a literary purpose. But the frequency of embarrassing *truths* that also serve a literary purpose is far from 100 percent. Most such truths will not "coincidentally" be convenient to tell—this is, in fact, the assumption on which EC reasoning is based, so if it weren't true, *no* EC argument would be valid.

We next must ask, how many embarrassing truths were there, in ratio to embarrassing myths that were found useful to tell? Since we don't know the answer to that question (certainly not *a priori*), not even whether there were more or fewer of either, the principle of indifference entails we cannot assume there was more of either (again, until we can prove otherwise). From these facts the conclusion follows: the frequency of embarrassing stories preserved that are true (among all embarrassing stories) will be low, not high. Hence, most such stories will be false. It's simple math: Even starting with equal numbers of each—say, ten and ten—if few of the former will be preserved but all of the latter will—let's say, only two in the first case but all ten in the second—then the prior probability that any surviving embarrassing story is true will always be low—in this case, two out of twelve (two plus ten equals twelve embarrassing stories preserved, of which only two are true), which equals one in six, less than 17 percent.

This would not be the case when a neutral or hostile tradition exists that is highly reliable and well known, such that a friendly source cannot easily ignore it or change it. Thus, if we can show an embarrassing claim had already been reliably and widely established by a neutral or hostile source by the time a friendly author wrote, then we can reverse this probability, because the frequency of *those* claims that will be true will be much greater. This does not follow when a neutral or hostile tradition exists that is not highly reliable (or dates *after* a friendly source and merely responds to it), because such traditions have their own tendency to fabricate embarrassing stories about a subject (to which a friendly source can respond by a variety of strategies, not just gainsaying or refutation, but even continued fabrication—because, after all, the friendly source might no more know the truth of the matter than their opponent did). But when it comes to the historical Jesus, we have no neutral or hostile sources of any kind (apart from much later critics who had no access to any information about Jesus not provided by the Christians themselves), and of any such traditions that *might* have existed before the Gospels were written (such as any we might try to infer from the Gospels or Epistles), we can establish none as reliable, early, or widely known. So we're back to the original probability.

Thus, for the Gospels, we're faced with the following logic. If $N(T)$ = the number of true embarrassing stories there actually were in any friendly source, $N(\sim T)$ = the number of false embarrassing stories that were fabricated by friendly sources, $N(T.M)$ = the number of true embarrassing stories coinciding with a motive for friendly sources to preserve them that was sufficient to cause them to be preserved, $N(\sim T.M)$ = the number of false embarrassing stories (fabricated by friendly sources) coinciding with a motive for friendly sources to preserve them that was sufficient to cause them to be preserved, and $N(P)$ = the number of embarrassing stories that were preserved (both true and fabricated), then $N(P) = N(T.M) + N(\sim T.M)$, and $P(T|P)$, the frequency of true stories among all embarrassing stories preserved, $= N(T.M) / N(P)$, which entails $P(T|P) = N(T.M) / (N(T.M) + N(\sim T.M))$.[80] Since all we have are friendly sources that have no independently confirmed reliability, and no confirmed evidence of there ever being any reliable neutral or hostile sources, it further follows that $N(T.M) = q \times N(T)$, where $q \ll 1$, and $N(\sim T.M) = 1 \times N(\sim T)$: because all false stories created by friendly sources have motives sufficient to preserve them (since that same motive is what created them in the first place), whereas this is not the case for true stories that are embarrassing, for few such stories so conveniently come with sufficient motives to preserve them (as the entire logic of the EC argument requires). So the frequency of the former must be 1, and the frequency of latter (i.e., q) must be $\ll 1$. Therefore:

$$\frac{N(T.M)}{N(T.M) + N(\sim T.M)} = \frac{q \times N(T)}{[\,q \times N(T)\,] + [\,1 \times N(\sim T)\,]}$$

$$= \frac{q \times N(T)}{[\,q \times N(T)\,] + N(\sim T)}$$

So even if we substitute 0.5 for q (which is much too high, as it would entail that half of all embarrassing truths brought no sufficient motive to suppress them, directly contradicting the basic assumption of an EC argument), this produces:

$$\frac{0.5 \times N(T)}{[0.5 \times N(T)] + N(\sim T)}$$

So if $N(T) = N(\sim T)$, then:

$$P(T|P) = \frac{0.5N}{0.5N + N} = \frac{0.5N}{1.5N} = 0.333$$

Or in other words, the number of embarrassing stories preserved that are true will be no greater than 1 in 3, which means that at least 2 in 3 (in other words, most) will be false. If for q (the proportion of all embarrassing truths that brought no sufficient motive to suppress them) we substitute 0.25 instead of 0.5, then these final fractions become 1 in 6 and 5 in 6, respectively—and 0.25 is already unbelievably high. 1 in 4 embarrassing truths are not convenient to tell. Surely more than 3 in 4 such truths entail a sufficient motive for a friendly source to suppress or forget them. If we concluded, instead, that at least 9 in 10 would, which is more credible (and advocates of the EC have certainly behaved as if the frequency must have been that high, for them to express such certainty that preserved embarrassments must be true), then the fraction of surviving embarrassing claims that will be true would be 1 in 11 (or only 9%). And one might even suspect *that's* too high. Is it really unreasonable to think that less than 1 in 100 true embarrassing stories are going to be "convenient to tell"?

Thus as $N(T.M) \rightarrow 0$, so does $P(T|P)$. Only if $N(T) >> N(\sim T)$ do we get $P(T|P) > 0.5$. Which means $N(T)$ must be greater than $N(\sim T)$ by as many times as 1 is greater than q.[81] In other words, the more frequently *in general* that embarrassing truths have sufficient motives to suppress them, the more embarrassing truths about Jesus there must have been than embarrassing but useful myths about Jesus in order for most of the embarrassing stories preserved to be true. And yet we know, more likely than not (and certainly so far as we know, absent circular logic), $N(T) \approx N(\sim T)$. In fact, it's very possible that $N(T) << N(\sim T)$, since one can fabricate countless embarrassing but useful myths, whereas there are only so many embarrassing things that can *actually* have happened to someone. That would make the prior probability that embarrassing stories in the Gospels are true *extremely* small. But no matter what the actual ratio, surely the number of useful embarrassing myths that can be conceived is always far greater than the number of actual embarrassing things that could have happened. So in no case is it ever likely that $N(T) > N(\sim T)$.

To make any headway toward reversing this judgment you must advance some theory as to why, despite all this, the originating author was compelled to include that detail anyway (if it were indeed known to be true and embarrassing to them), and then from that theory you must estimate how compelled the author was. From this you can derive a narrower reference class from which to revise the prior (on which see chapter 6, page 231) or generate a difference in consequents that overcomes the prior. But you can't just make up any reason willy nilly. Because if you can't prove that that reason was actually operating on that author, or even likely to be, then your theory's initial probability will be lower, not higher (as the probability space must then be evenly divided among several competing theories, representing all the other motives just as likely operating on that author instead of yours, as explained for all *ad hoc* reasoning in chapter 3, page 80). That means your prior probability (which, as just argued above, is in this case *already* low) must be adjusted even further downward to reflect this fact. On the other hand, any reason you can demonstrate *was* operating on that author, or very likely to have been, will raise the prior only if it makes that author more likely to report that fact if true than to fabricate it if useful. Of course, whatever reason we demonstrate an author had must entail that the strength of his impulse to include a detail despite its embarrassing nature was *stronger* than any impulse he had to omit it, as that's the only condition in which an embarrassing statement will be included against an author's interest. And that impulse must be strongly correlated with that statement being true (as otherwise an author may feel compelled to include it *thinking* it's true, when in fact it's not—hence, we still must attend to the second prong of a proper EC test).

So unless we can demonstrate otherwise in any particular case, or unless we can demonstrate an author to be remarkably honest and reliable in making claims about Jesus generally, the prior probability is low that an embarrassing story in the Gospels is true. Most of those stories will be in them because they were useful to create; few will be in them because they were true. Then you must estimate the consequent probabilities of the evidence given your competing theories (which theories are that an author or source had reason to fabricate the statement vs. that an author made that statement only because it was true). How likely is it that *that* detail would have been included by *that* author, in *that* text, in *that* context, and in *that* way, given that it's true? The overall evidence might not fit that hypothesis so well. If, for example, we should expect an embarrassing fact to be reported only in conjunction with some sort of apologetic, yet it's not, then P(e|h.b) is low. For on the hypothesis that an author reported something he didn't want to report because he was compelled to for some reason, odds are the author would express this, or in some fashion explicitly defend his interests against the implications of the embarrassment, which means if this didn't happen, then the evidence is not as expected on our hypothesis.

Still, in some cases the evidence can fit our hypothesis fine, being entirely what we should expect (entailing P(e|h.b)→1). The remaining problem is that in all the cases we find in the Gospels, there are also good theories about why Mark (or his sources) would fabricate the allegedly embarrassing detail, which entails P(e|~h.b)→1. So even finding a version of a story in Mark exactly as we should expect it to appear (with evidence of his embarrassment, let's say), it still might not be credible. But usually such "evidence of embarrassment" won't be very explicable on any alternative explanation, and thus the consequent probabilities will favor an embarrassing story's being true in that case—*if* we can establish that Mark probably believed it *because it was true*, rather than for some other reason, yet Mark never tells us anything about his methods or sources or the warrants for his beliefs, nor can we securely deduce any of these things. Hence, it seems we are always stymied by one or the other prong of a proper EC criterion: for any given claim, either we can't show it was against Mark's interests to claim it, or we can't show he would have reliably known it was true (whether it embarrassed him or not). And as for Mark, so for his sources.

But the main obstacle in Jesus studies is that strong plausibility of alternative explanations. Thus Craig Evans's attempt to reformulate the EC as "authenticity is supported when the tradition cannot easily be explained as the creation of the Church in general" at least correctly acknowledges the importance of establishing a low $P(e|\sim h.b)$.[82] This is even a nice example of a coded phrase in English ("cannot easily be explained") representing what is in fact mathematical reasoning: that the consequent probability for $\sim h$ (that the material in e is "the creation of the Church in general") must be low. But establishing such in the case of the Gospels is a lot harder than most historians casually assume. And the consequent probability for h (that the material in e is "the product of having actually happened") must still be high (which it often is not, e.g., the Judas story, which makes no sense as actual history—and that, being highly contrary to expectation on any hypothesis that it actually happened, entails a low $P(e|h.b)$).

Often the prior probabilities must favor h over $\sim h$ as well, which requires (among other things) the prior demonstration that the Gospels are honest historical records rather than deliberately constructed myths. Such a prior demonstration must conform to BT and could logically consist of beginning with neutral priors (50/50, favoring neither theory about the nature of the text) and then running case after case until the trend is visible (either mostly history, mostly myth, or equal parts either), the outcome of each case becoming the prior probability in the next. That trend will then equal the prior probability of meeting that same trend again in the next case, the result of which will then become the prior probability for the *next* case, and so on (this is a mathematically valid procedure—which I'll discuss again in chapter 6, page 239). Absent such a proper demonstration, as previously shown, the priors will favor fabrication in every case of seeming embarrassment. In fact, for embarrassing elements, that genre-based prior would normally be your initial prior in any such analysis (not a blind 50/50 probability), which means this process of iterating the analysis case after case would have to repeatedly and strongly favor truth-telling to get that prior probability up to any respectable value in the case of the Gospels, just as with any other sacred stories for any other religion.

Either way, what we are still faced with is a question of a balance of three different probabilities. We can then use BT to produce a conclusion. But since, as a result of all the above, the conditions for a successful EC argument are so rarely encountered, and probably never encountered in the Gospels, these probabilities never favor historicity enough, so the Criterion of Embarrassment can have little or no use in reconstructing facts about the historical Jesus.

COHERENCE

All the other criteria suffer the same defects. The Criterion of Coherence assumes that anything that coheres with what has been established with other criteria is also historical. "If a saying or story is consistent with the picture of Jesus that is emerging from well-attested material, then it may be considered likely to have been historical," because "it is consistent with what has already been discerned" as originating with Jesus.[83] But this is illogical.[84] Coherent material can be fabricated precisely *because* it coheres with other beliefs about Jesus, or even for the specific *purpose* of cohering with them.

Liars tend to prefer their lies to be coherent, and when telling new lies, build on old ones. Even more innocent mechanisms of legendary development follow the same principles ("but that's just what Jesus *would* have done, isn't it?"). Indeed, the usual trend in fabrication over time is toward harmonization—in other words, the *creation* of coherence (the textual tradition of the Gospels even shows this explicitly occurring). Everyone knows "good fiction is often just as 'coherent' as historical fact."[85] Indeed it can even be more so—for coherence is easy to create by design, whereas real historical people and events are often evolving, complex, unpredictable, or actually in fact incoherent. It's easy not to understand what people said and did, to see such events as incoherent or inexplicable. It's then much easier to make it all cohere *after the fact*, imposing structure and consistency on what actually had none (or whose structure and consistency actually escaped you, and thus was replaced by structure and consistency of your own imagining). As a result, although *failing* to cohere with established facts might (at least in some cases) raise the priors or consequents of "fabrication" hypotheses, *cohering* with established facts does not raise the priors or consequents of hypotheses of "historicity" by any relevant amount—since coherence is just as common and expected on hypotheses of fabrication.

As Anthony Le Donne aptly puts it, this criterion suffers from a one-two punch: "it presupposes that certain characteristics of Jesus are of little dispute" yet "such characteristics are very few," and even those are suspect (as is becoming clear throughout this chapter—and as even Le Donne himself warns, "this criterion has a tendency to confirm the presuppositions of the scholar" rather than any actual facts); and even when granting a characteristic as independently established, "it is possible that a characteristic well known to early memory may have bled into other episodes during narration" and thus reflect *accidental fabrication*, not historicity.[86] Add to that the occurrence of intentional fabrication and *its* tendency to cohere, and coherence just isn't a reliable indicator of the truth.

It might be assumed that prior confirmed cases of some property x would increase the prior probability of another case of x being true, but in this case that doesn't follow. A story found in an author whose stories often turn up confirmed *will* have a higher prior probability of being true; but the mere fact of "a" detail cohering with "some other" detail in the overall "tradition history" of all the stories preserved does not have that effect, precisely because not all authors are reliable, nor are all sources, and unreliable authors and sources will produce (or reproduce) fictions that cohere with tradition more than amply. Thus, the *mere* fact of cohering with established facts is insufficient to make a story more likely to be true. If, for example, unreliable sources transmit as many false cohering stories as reliable sources do true ones, and we cannot confirm there are more reliable sources behind our accounts than unreliable ones, then so far as we know the set of all cohering stories will contain as many fictions as truths, making no difference to the odds that any particular story will be true. Only if you can demonstrate that these ratios fall out differently can you get coherence to change the odds of some detail being true. And in the study of Jesus, we generally don't have that kind of information.

Even at best, in every case you have to attend very carefully to the evidence. For example, it's sometimes assumed that Matthew 10:5–6 and 15:24 (where Jesus says only Jews are to be evangelized) must be authentic because "we know" that Jesus didn't call for any mission to the Gentiles (because according to Galatians that mission was initiated by Paul after, as

Paul himself says, Jesus had died). But there might be no reason to assume Jesus ever had cause to *tell* anyone not to go into Gentile communities to evangelize them (why would that even occur to his disciples and thus have to be explicitly prohibited in the first place? And even if it occurred to them, why would they go do such a thing on their own initiative, rather than await his specific orders?). In contrast, we know (from the way he redacted Mark) that the writer of Matthew is up against a Gentile Christian community he wants to discredit, so he clearly had motive to invent these sayings for Jesus, to further his agenda.[87] So the fact that they cohere with what may be an established fact about Jesus does not lend any weight to these sayings being historical. To the contrary, the evidence suggests they probably aren't (e.g., we can see Paul never had to answer anyone quoting these sayings of the Lord against him, yet surely they would have, so the consequent probability of all the evidence together is actually *higher* on the hypothesis of fabrication). On the other hand, Matthew's seemingly contradictory *endorsement* of a mission to the Gentiles (in Matthew 28:19) is no more likely to be *unhistorical* because it *fails* to cohere with what we know from Galatians—because Matthew does not mean what Paul was doing (converting Gentiles straightaway, without first converting them to Judaism through circumcision and dietary laws), but what we know Jesus' disciples were already doing even before Paul came along (allowing Gentiles who become Jews to join the Christian community, as is clearly attested in Galatians they had already done before Paul and were ready to continue doing), which is what many Jews in antiquity did. The idea of evangelistic Jews sounds odd today, but only after thousands of years of anti-Semitic legislation and hostility has chastened Jews into giving up the active pursuit of converts as being bad for their health. But in antiquity they were often welcoming or even pursuing Gentile converts.[88] It's entirely possible Jesus preached the same practice (it was, after all, God's law [Exodus 12:48], "not one jot or tittle" of which Jesus intended to abolish, according to Matthew 5:17–18). Because if he did, that would in no way fail to cohere with the evidence in Galatians where the Apostles before Paul were doing this very thing already; and certainly Matthew consistently imagined such a mission throughout his Gospel (Matthew 24:14,

25:31–46). Of course, all of that still doesn't mean any of these sayings are or aren't historical, only that we can't rule them in or out using the Criterion of Coherence, as many scholars have attempted to do.

And that's even assuming you actually have established facts to cohere with (or not). The greatest folly in applying this criterion is the same bootstrapping fallacy critiqued earlier: "cohering" with a "fact" established by an invalid (or invalidly applied) criterion cannot legitimate another fact. Yet most "historical Jesuses" are constructed from exactly such a house of cards. Hence, the Criterion of Coherence is the most insidious of them all.

MULTIPLE ATTESTATION

"If a tradition is attested in more than one strand of the tradition" then "it is more likely to be authentic," as long as these strands are "*independent* layers of the tradition."[89] This is often a sound principle commonly employed throughout the field of history. But it has to be applicable—and applied correctly. And yet in Jesus studies, "relatively few individual units of the tradition are attested in more than one strand," and even in those few cases establishing independence is hard to do.[90] It was long thought that the Gospel of John is independent of the Synoptics, but a growing body of evidence argues otherwise, so John's independence can no longer be reliably assumed.[91] Some scholars even argue that Mark knew Q (which is entirely possible: just because he rejected much of it doesn't mean he wasn't aware of it or didn't use it) or that there *was* no Q, only Matthew's expansion of Mark.[92] And hardly any of the extrabiblical evidence for Jesus is independent of the NT [New Testament], and most of the evidence that even so much as *might* be independent of the NT is universally rejected as fabricated (e.g., the Infancy Gospels; 3 Corinthians; the Epistle of Jesus to Abgar), and thus can hardly count as "multiple attestation" (a fact that should already caution against assuming the canonical texts are any more trustworthy than these).[93]

Those facts put any reliance on the Criterion of Multiple Attestation on very shaky ground, particularly for the reasons just noted for the Criterion of Coherence: we should actually *expect* multiple attestations to be fabricated. Hence the Infancy Gospels "corroborate" that Jesus was a great miracle worker, yet we know full well this evidence is fictional—and thus doesn't corroborate anything. And even when we have something like a credible instance of independent attestation, that only proves that the corroborated datum originates in an earlier source, *not* that it originates with a historical fact. Because "multiple attestation . . . does not exclude the possibility of creation by Christians" at earlier stages of development prior to the documents we have.[94]

A classic example is the fact that we have impressive multiple attestation of the labors of Hercules (and in antiquity this was even more the case, as many texts now lost are known to have recounted them), yet no one believes this makes those labors even "more probable," much less believable. The story was fabricated long before our written sources (probably long before even *their* written sources), spawning numerous independent lines of legendary development, which each came to be independently recorded later on, an outcome that in no way makes the story more credible. The same clearly happened to the Christian tradition before the Gospels were even written. For example, it appears two separate and contradictory legends developed about the suicide of Judas (Matthew 27:3–10 vs. Acts 1:18–20), yet this could have resulted just as easily from an originating fiction as from an originating fact (and as I argued earlier, it most likely did: see page 152). Thus its multiple attestation does not establish its historicity (and if Luke's version is a deliberate rewrite of Matthew's, we don't even have multiple attestation). A similar datum can also originate independently because of a common *motive* rather than a common *source*, to explain a shared problem in the text or to defend a shared doctrine or goal. For example, many scribal emendations of the NT manuscripts produce the same or similar results even when they were not even aware of each other, simply because they saw the same problems and devised similarly obvious solutions. Though this will usually be less likely, it still has to be ruled out.

For all these reasons it's simply not true that, as Marcus Borg claims, "if a saying or story appears at least twice in traditions that are early and independent of each other, that is a very good reason for thinking that the gist of it goes back to Jesus."[95] Because it is an equally good reason for thinking that the gist of it goes back to an originating myth (or even a revelatory dream or vision), or an earlier storyteller's innovation. For example, Borg's own example of the general fact of Jesus performing "healings and exorcisms" could just as easily be an invention to idealize and legitimize the fact that early Christian communities were engaging in healings and exorcisms. After all, Paul attests to such activities in the earliest churches, yet never attests to such things ever having been done by Jesus (try as you might, you won't find the latter in his letters, but you'll find many hints of the former). If a myth grew during this time that Jesus did them, too, it could easily have spread to independently inspire different stories in the Gospels, just like the labors of Hercules. And that's not the only possibility. For if Q is actually just the elaborations on Mark produced by Matthew (as Goodacre argues—remember, Q is entirely hypothetical, it is not a document we actually have), then Jesus being a healer and exorcist could have been an outright invention of *Mark*, which thereby inspired all subsequent iterations of the theme—because then it does not appear even twice in traditions "that are early and independent of each other," but only in Mark, and in documents long post-dating and even employing Mark. The fact is, these questions are not as settled as Jesus historians often claim.

When does Multiple Attestation argue for historicity? Only when the fact of multiple attestation entails $P(e|h.b)$ is substantially higher than $P(e|\sim h.b)$. Which only occurs when we can establish (to some degree of probability) that two extant testimonies to the same claim derive (at least ultimately, if not directly) from *independent eyewitnesses* of the fact attested, or from one eyewitness (or group of eyewitnesses) whose reliability on that claim is demonstrably more likely than their lying or being in error. But rarely can we ascertain even who an author's source is, much less to which eyewitness it can ultimately be traced, and we can rarely assert someone is reliable when we don't even know who they are. Even when we do know, it would be naive to merely *presume* their reliability; and establishing it is often impossible. Which is why hearsay is almost never even *admitted* as evidence in a court of law, and why modern historians of antiquity are often skeptical of all but the most public or mundane of claimed "facts."[96]

Notably, the most persuasive cases of multiple attestation are when our sources are diverse in type—as exemplified in my analysis of the Argument from Evidence in chapter 4 (which is precisely why that method is so useful: see page 98). But this is exactly what we don't have in the case of Jesus. Rather than items from all five categories of useful evidence, we don't even have one of them. All we have are uncritical pro-Christian devotional or hagiographic texts filled with dubious claims written decades after the fact by authors who never tell us their methods or sources. Multiple Attestation can never gain traction on such a horrid body of evidence.

EXPLANATORY CREDIBILITY

Any claim that "seeks to provide a plausible explanation for the rise of Christianity within its first-century Jewish context" can be considered a candidate for historicity.[97] Stated thus it can only be exclusionary (claims that fail to meet this criterion are probably false, but claims that meet it are not thereby true), and is only valid insofar as it merely repeats BT, i.e., a "plausible explanation" can mean an explanation with a high prior, and yet explanations with lower priors can actually be more credible if their consequents are sufficiently higher; conversely, a "plausible explanation" can mean an explanation that makes the evidence we have more probable than other explanations do, which is merely an assertion of higher consequent probability, and yet explanations with lower consequents can actually be more credible if their *priors* are sufficiently higher. So this criterion itself is useless. Only when we replace it with BT do we get something valid to work with. This same analysis follows for *all* other exclusionary criteria (below).

CONTEXTUAL PLAUSIBILITY

To be historical, Jesus must have been a Jew in early first-century Judaea, which had been gradually Hellenized for centuries (through conquest, immigration, and trade) and was at that time infiltrated, influenced, and governed by Romans. So "anything Jesus said" or did (or that was said or done to him) "must therefore make sense within the religious, social, cultural and linguistic milieu of that context."[98] This is again only exclusionary (claims that fail to meet this criterion are probably false, but claims that meet it are not thereby true). And to that extent it merely states that anachronistic claims about Jesus have a low P(e|h.b). Whether that means such claims are false still requires working out the rest of the equation; and whether claims that *aren't* anachronistic are *true* requires doing the same. Because as suggested earlier, good fiction will be contextually accurate, especially if it originated early—and sometimes even if originating late, as there were then many books available providing accurate historical detail for a period and place, which a storyteller could employ as references to lend veracity to his tales.[99] And often many details remained alive in living memory or were actually commonplace and thus not as distinctive of the particular time and place as might be assumed. Thus getting such details right does not automatically reduce P(e|~h.b). Though that's the assumption this criterion is sometimes meant to capture, it only sometimes has that result, requiring more analysis than this criterion alone entails. Conversely, it's not automatically the case that an anachronism is false. We may be wrong about whether a given detail is in fact anachronistic, so of course we can't circularly assume it is, and occasionally unusual behaviors do occur ahead of their time. Thus, demonstrations are still needed, and BT is the only method up to the task.

HISTORICAL PLAUSIBILITY

"Any reconstruction of Jesus must show that it is 'historically plausible' in the widest sense of the phrase: it must cohere with, and make sense of, all the evidence we have," which also means, "his life and teaching must be such that the written accounts which eventually emerged are explicable."[100] But this is either invalid or merely restates BT and thus should simply be replaced by it.[101] The first statement is basically just BT in a nutshell, for only BT validly measures the degree to which any theory "coheres with, and makes sense of, all the evidence we have" (the former referring to prior probability, the latter to consequent probability). The second statement essentially just insists we need a high $P(e|h.b)$. But that alone does not entail $P(h|e.b)$ is high, nor does a low $P(e|h.b)$ entail $P(h|e.b)$ is low; to know one way or the other you have to work out the rest of the equation. Thus BT supersedes this criterion.

NATURAL PROBABILITY

Claims contrary to nature or that are suspiciously improbable are probably false.[102] This is essentially just a restatement of the Smell Test analyzed in chapter 4 (and there demonstrated to be yet another special case of BT: see page 114). It's another criterion that is primarily exclusionary: validly applied, it can only tell you what's probably false, not what's probably true—except sometimes when a naturalistically realistic claim is made that an author could not have imagined with the same frequency as someone actually having seen it. For example, when Pliny the Elder marvels at a tribe of fire walkers, it's unlikely he or anyone would have made this up had it not been true. For that would entail a remarkable coincidence between random fantasy and an actual, venerable, scientifically well-understood magic trick (the details of which his account perfectly corresponds to). Likewise, Pliny the Younger's once scientifically incredible description of the eruption of Mount Vesuvius was scientifically confirmed by the modern eruption of the geologically similar Mount St. Helens, which vindicated every detail. It's quite unlikely the younger Pliny would have just accidentally erred his way into what turned out to be a scientifically correct observation.

Unfortunately, this reasoning does not have much use in the study of Jesus, whose only naturalistically credible miracles (such as psychosomatic healings and exorcisms) were commonly practiced by subsequent Christians and thus could have been fabricated quite easily by simply projecting onto Jesus the practices of Christian communities (thus legitimating them and creating models for them). Even his failure to succeed in his own hometown (according to Mark 6:1–12) has the same ready explanation. This is sometimes explained as evidence his power was indeed only psychosomatic, which by extension proves the historicity of his role as a faith healer. But that doesn't follow. Because it just as easily proves fiction, by again projecting onto Jesus the practices of Christian communities (thus legitimating them and creating models for them), in this case the occasionally inevitable failure of Christian faith healers, similarly explained as a lack of faith among the sick—not only the same excuse used by faith healers even now, but the very excuse used in Mark 6:6, thus establishing another useful model and precedent (leaving Christians the convenient justification "even the Son of God could not heal the unbelieving"). This pericope also addressed the occasional problem of Christian missionaries facing conflict from their families; hence, another obvious point of Mark's story would be to manufacture a context in which to coin the proverb in Mark 6:4. Such useful fiction is actually the better explanation, as it would be naturalistically improbable that a psychosomatic faith healing act would fail like this in only the one town (as if that were the only repository of the unbelieving). But a story created for its utility would not need more than the one event. Thus the consequents favor fiction. So again BT prevails over the criterion alone.

ORAL PRESERVABILITY

"If we cannot imagine a tradition being preserved orally, then we cannot think that it goes back to Jesus," such as, Marcus Borg argues, "the extended discourses attributed to Jesus in John's gospel," because "what one can imagine being remembered is the gist of a saying, parable, or story," whereas it's hard to "imagine one or more disciples memorizing [e.g., the entirety of John 14–17] as it was spoken and then preserving it through the decades" (in fact nearly a century, according to some scholars), especially with no other Gospel author in the meantime ever having heard of it.[103]

Again this criterion has no positive heuristic value (being exclusionary, it cannot tell us what is authentic, only what isn't), but it does reflect a valid historical intuition: such extensive speeches (as well as—and it's important to repeat this—every other author's ignorance of them) are improbable on a hypothesis of "accurate oral memory," but highly *probable* on a hypothesis of "fabricated to the occasion to sell the creedal ideas of the authors of John." Hence, the balance of consequents favors fabrication. The prior probability of such an amazingly accurate oral tradition is *also* extremely low for the first-century Christian tradition specifically. Our background knowledge establishes that oral tradition can be rapidly distorted and expanded with fictions, while any *true* content gets simpler and less detailed over time—indeed, writing was specifically invented to combat those facts (and yet even writing suffers alteration and distortion over time, just much more slowly), which facts are confirmed by recent scientific studies of human memory.[104] Only when background knowledge establishes a high prior can the inevitably low consequent be overcome, and we can't do that for early Christian oral tradition. We have no evidence of the presence and operation of the institutions and mechanisms we know are required for securing the reliable memorization of detailed material within the first-century churches (there were no Christian schools, for example, as there were for memorizing the Mishnah, nor was the Gospel put into verse, song, or anything like mnemonics or counting rhyme). Indeed, the wild discordance among the New Testament materials (much less noncanonical materials) attests no such institution or mechanism was in operation. Thus Borg's intuition is sound. And again it's BT that proves it.

CRUCIFIXION

"Any proposed reconstruction of Jesus has to be a Jesus who was so offensive to at least some of his contemporaries that he was crucified."[105] In other words, he must have committed some outrageous crime or posed some very real threat, otherwise no one would have bothered. But this assumes *a priori* that he was crucified (which also, of course, assumes he existed). Reformulating the criterion so as not to beg the question gives us "any proposed theory of Jesus has to explain why he was preached crucified." Analogously, "any proposed theory of Attis has to explain why he was preached castrated," "any proposed theory of Inanna has to explain why she was preached humiliated," and so on. All true. In formal terms, this amounts to asserting that any h that doesn't make reports of Jesus' crucifixion likely will entail a low $P(e|h.b)$, *and* a high $P(e|{\sim}h.b)$ since many other hypotheses make those reports likely, thus the consequents will weigh heavily against your theory. But e must include the Gospel accounts of the crucifixion, which already substantially lower *both* consequents for any theory that isn't elaborately complicated, which theories often have low priors (because increasing a theory's complexity decreases its prior: see chapter 3, page 80). For the explanation given in the Gospels makes little to no historical sense (as remarked earlier, page 140). The *historicity* of the crucifixion of Jesus is thus as challenged as any theory positing it as myth. Either way some elaborate *ad hoc* theory is required to explain all the oddities of the evidence.

Thus this criterion is useless. You have to fall back on BT. In other words, though the premise seems correct (any theory you have must make "Jesus was preached crucified" likely), there are many more ways to meet that premise than normally assumed (the number and diversity of theories of comparable prior probability is great), and even not meeting it does not automatically entail a theory is untrue (since a large enough disparity in priors can overcome any disparity in consequents, i.e., sometimes unlikely things *do* happen).

FABRICATORY TREND

"Whenever a saying or story reflects a known tendency of the developing tradition, the historian must suspect that saying's historicity," which is again only exclusionary.[106] For example, the trend in the Gospels toward magnifying Jesus over time casts such magnification into doubt. And yet Paul, our earliest source, already has a rather high Christology, substantially predating the Gospels. So you have to take great care to ensure the trend you claim is there is really there, and not something that's been inherent in the tradition from its inception. But once that requirement is met, this criterion becomes a valid element of determining prior probability. In effect, to say that a claim conforms to an observed trend of legendary development is to say that our background knowledge establishes a high prior probability that more of the same will also be legendary. Of course, that prior being high does not alone entail such a detail *is* legendary, since a ratio of consequents strongly favoring historicity can still prevail. Hence, this criterion is again subsumed by BT.

LEAST DISTINCTIVENESS

This is the assumption that when we have many versions of a story or saying, the simpler and less elaborated is the earlier and thus most authentic (in accord with the Criterion of Oral Preservability above, page 178). This is actually just a special case of the previous criterion (Fabricatory Trend), yet one that's multiply problematic. For example, "traditions becoming longer and more detailed, the elimination of Semitisms," or "the use of direct discourse, and the conflation and hence growth of traditions" are different ways a simpler account becomes elaborated over time.[107] And yet, as Stanley Porter observes, sometimes the less elaborate version is an edit or truncation of an earlier, more elaborate version, and sometimes Semitisms are *added* and thus a result of embellishment, not a sign of being more original. Moreover, an earlier, simpler version is not thereby true. Even the presence of early Semitisms does not make a story more true (as we'll see under the Criterion of Aramaic Context, below, page 185).

So this criterion is at best only exclusionary, and not universally applicable. But it could be helpful. If the evidence in a specific case *is* such that we can prove (to some degree of probability) that an elaborate version is more probable as an embellishment of the simpler version than as the retention of an earlier, truer account (or the simpler version is *less* probable as a truncation of the elaborate version than as a retention of an earlier, truer account), then we can say the consequent probability of the simpler version being true exceeds the consequent probability of the elaborate version being true. But that would entirely depend on cases. And again, the priors can reverse the outcome. Or indeed so can the consequents: if the evidence actually supports the contrary conclusion that the simpler version is the more derivative, or verifies one as the earlier but fails to establish that it's thereby true—because even the earliest, simplest version of a story we have is still often nevertheless a fabrication. So BT must still prevail.

More interesting is how this criterion could affect the prior probability. Rather than simply asserting that the criterion is true, historians must actually test it: examine every case where we have simple and elaborate versions of a story or saying, starting with a neutral prior (0.5) and using the outcome of each case as the new prior in the next (the procedure suggested earlier, page 168). If after numerous representative cases the trend shows a higher prior probability confirming the trend (i.e., in most simple-elaborate pairings, the elaborate version is indeed the less credible), then you can apply that prior to further cases (continuing the same procedure). And of course, doing this may get you to the earlier versions of stories, but not all the way to knowing whether those stories are at all true, which may require further analysis (if more even can be known on that point). So once again this criterion is only valid when replaced by BT.

VIVIDNESS OF NARRATION

As if defiantly contradicting the previous criterion, it has also been claimed that versions of a story that are more vividly narrated (as if the author were "there" and viscerally responding to what she experienced) are more likely true than versions that use a more distant or cursory mode of narration. But this is a non sequitur. For vivid detail is *also* an established trend in fictionalization and embellishment. Good storytellers often come up with these details, especially when they are lacking, and thus such elements are as likely as any to accumulate in the retelling over time. Human memory even does this routinely, without anyone being aware of it, especially through memory distortion and contamination through repeated retelling.[108] Conversely, an eyewitness can produce a concise and droll account without vivid narration, as can someone relaying what an eyewitness said.

In fact, in the ancient world especially, our background knowledge establishes that vividness of narration is more often a sign of fiction than history. Schools of the time specifically taught writers to embellish stories and speeches in exactly this way (see note 118, page 323), and we can find numerous cases where battles and speeches are described in vivid detail when we know for a fact the author had no actual sources for them. Conversely, the histories we trust the most are the ones that restrict themselves to the fewest and least embellished details that could be corroborated by multiple lines of evidence, and we especially prefer prosaically analytical discussions of the evidence to unsourced novelizations of it.[109] Thus, vividness of narration by itself actually argues slightly *against* historicity, not in favor of it, and can only support historicity in cases where you can specifically prove such vividness isn't the result of dramatization. So by itself this criterion is useless, and we're left again with BT.

TEXTUAL VARIANCE

Formally, this is called the "Criterion of Greek Textual Variance." According to Stanley Porter, "where there are two or more independent traditions with similar wording, the level of variation is greater the further one is removed from the common source," whereas the presence of "less variation points to stability and probable preservation of the tradition, and hence the possibility that the source is authentic to Jesus."[110] But there is actually no basis in our background knowledge for either view.

The stability of a tradition no more demonstrates its authenticity than its later importance or familiarity—or in some cases even its *unimportance* (insofar as passages more doctrinally charged received the most meddling). Sometimes variance decreases with fatigue: innovation in adapting a source decreases as an author goes along until that author more lazily just copies his source.[111] Yet that would not indicate the latter half was any more true than the former. Even in the best of cases one might use this criterion to be more certain of what an author's immediate source said (hence this criterion is a fundamental tool in textual criticism), but that doesn't help us much, since what a source said and what's historically true are not thereby identical. Certainly the transmitters often can't possibly have known whether what they were transmitting was *really* true or not, so their passing it on more consistently can't be in consequence of it being true. Just as often it will have other causes (such as any of the four just enumerated). In fact, that some traditions will show greater or lesser variation is expected simply as a consequence of completely random fluctuation. To then point to the ones that just by chance got the less and claim they are more true is logically perverse. Either way this criterion has no validity.

GREEK CONTEXT

Formally, this is called the "Criterion of Greek Language and Its Context." According to Stanley Porter, "if there are definable and characteristic features of various episodes that point to a Greek-language based unity between the participants, the events depicted, and concepts discussed," then "the probability would be greater that Greek would have been the language of communication used by Jesus and his conversation partners" and therefore it "might well have originated with Jesus."[112] Unfortunately this is a cascade of non sequiturs. For one can just as easily argue from such a fact (assuming it's even established) that this conversation was entirely constructed by the author, or by another Greek-speaking source, and not derived from Jesus at all, which is even the more probable if you reject Porter's controversial theory that Jesus held conversations in Greek. Porter is also committing the fallacy here of bootstrapping a conclusion from "is more probable" to "probably is"—an invalid procedure (as proved earlier). This criterion has no validity.

ARAMAIC CONTEXT

Formally the "Criterion of Semitic Language Phenomena" (and known by other names), this is the flip-side of the previous criterion: if there is evidence of an "Aramaic-language based unity between the participants, the events depicted, and concepts discussed" underlying the extant Greek text, then this suggests the account goes back to the original Jesus, who most likely conversed in Aramaic.[113]

The first difficulty with this criterion is that it isn't easy to discern an "underlying Aramaic origin" from an author or source who simply wrote or spoke in a Semitized Greek. The output of both often look identical. And yet we know the earliest Christians routinely wrote and spoke in a Semitized Greek, and regularly employed (and were heavily influenced by) the Septuagint, which was written in a Semitized Greek. This is most notably the case for the author of Luke-Acts, and is evident even in Paul. Many early Christians were also bilingual (as Paul outright says he was), and thus often spoke and thought in Aramaic, and thus could easily have composed tales in Aramaic (orally or in lost written form) that were just as fabricated as anything else, which could then have been translated into Greek, either by the Gospel authors themselves or their sources. Indeed, some material may have *preceded* Jesus in Aramaic form (such as sayings and teachings, as we find collected at Qumran) that was later attributed to him with suitable adaptation. So even if we can distinguish what is merely a Semitic Greek dialect from a Greek translation of an Aramaic source (and we rarely can), that still does not establish that the Aramaic source reported a historical fact.

Consequently, Semitic features in a Gospel pericope do not make its historicity any more likely, other than in very exceptional cases (where we can actually prove an underlying source that we otherwise did not already suspect), and even then it gains very little (since an underlying source is not automatically reliable). Whereas one might have hoped such features would lower P(e|~h.b) relative to P(e|h.b), there is no evidence in *b* that warrants that conclusion. Even the best cases would lower it but little; and most cases, not at all. As Christopher Tuckett says:

> We should not forget that Jesus was not the only person in first-century Palestine; nor was he the only Aramaic speaker of his day. Hence such features in the tradition are not necessarily guaranteed as authentic: they might have originated in an early (or indeed later) Christian milieu within Palestine or in an Aramaic-speaking environment.[114]

Or as I've noted, they might have originated in a Semitic-Greek-speaking environment (of which there were many across the whole Roman world), or even a pre-Christian milieu. Even a chronological trend is not dispositive, since Stanley Porter finds evidence the tradition could become "both more and less Semitic."[115] Unfortunately there are just too many ways a Semitic flavor could have entered the tradition of any saying or tale, and we have no way to tease out their relative probabilities. So when it comes to Jesus, this criterion effectively has no value for discerning historically authentic material.

DISCOURSE FEATURES

Stanley Porter defines the Criterion of Discourse Features as follows:

> If the words of Jesus are determined to be significantly different from those of the surrounding Gospel, and especially if these words are consistent from one segment to another, then the presumption is that the author, and by extension any later redactors, of the Gospel have preserved the words of Jesus in an earlier form, ostensibly a form that could well be authentic, rather than redacting them as the Gospel was constructed and transmitted.[116]

This is unfortunately another non sequitur. Even if the procedure works (and in this case it hasn't been shown to), the most it could establish is that those discourses derive from a different source than the narrative material. That source will not necessarily be Jesus. It could be any storyteller. Just as discussed under the previous criterion, unless the Gospels were contrived from whole cloth, we already should expect a diversity of source materials in different languages and dialects. That does not make what they said true. Moreover, ancient authors were specifically taught to employ a different style in direct discourse than in descriptive narration, even to mimic (or create) distinctive styles for different speakers, and Porter's procedure can really establish no more than that they did that, so it probably can't even establish that a different source was used. So this criterion not only hasn't worked (the required procedure has never been attempted), it probably can't work.

Only if it was unexpectedly successful (e.g., we proved a consistent authorial style within the speeches of Jesus spanning all four Gospels, inconsistent with the style of all the remaining Gospel material) would we have a case for lowering $P(e|\sim h.b)$, since such a coincidence is unexpected on any theory but "common authorship." But since the common author need not be Jesus (it could still be a singularly influential missionary between Jesus and the extant Gospels, or even a teacher prior to Jesus whose teachings were later attributed to him), $P(e|\sim h.b)$ wouldn't necessarily be lowered by much. And this hypothetical outcome is not likely to be realized. For the discourse features of Jesus' speeches in John vs. the Synoptics are self-evidently not in agreement, while any agreement found between Mark and Q could be explained by Mark having selected his material from the same document (Q), or by Matthew and Luke emulating Mark in their adaptation of Q. With no way as yet to tell the difference, we have no valid use for this criterion.

CHARACTERISTIC JESUS

Finally, the most recent attempt at inventing a new criterion—on the heels of once again proving all the others defective—is the Criterion of the "Characteristic Jesus," which argues that "any feature that is characteristic within the Jesus tradition, even if only relatively distinctive of the Jesus tradition, is most likely to go back to Jesus."[117] As that is clearly a non sequitur (for all the reasons surveyed so far), this criterion is also invalid.

OTHER CRITERIA

Other criteria that I've seen implicit in various arguments are usually garden-variety fallacies. The notion that a story just sounds so real and moving it must be true I'm tempted to call the Criterion of It Just Feeling True. But logicians already named this years ago. It's called the Affective Fallacy: judging something true because of how it affects you (how real it sounds, how moving it is, etc.). Such reasoning has no objective merit. Another I call the Criterion of Inexplicability, which logicians have long identified as the Argument from Ignorance: the fallacy of assuming that because you can't think of any other reason a claim would exist, then it must be true; or assuming that because you can't find any specific evidence a claim is false, then it must be true. Neither assumption is logically valid. You must attend to questions of prior probability in light of the telltale features of the text in question (e.g., the Smell Test, examined in chapter 4), and whether you should even expect to have evidence against a claim if it is false (e.g., the Argument from Silence examined in chapter 4), and then weigh all this in a sound fashion (which is exactly what BT does), before deciding whether the absence of evidence against a claim actually warrants believing it. But usually any informed expert who takes the consideration of alternative hypotheses seriously will not only always find good contenders to test—she'll be able to think of several herself.

I've also seen what I call the Oral Source Fallacy: assuming that because you can't identify or reconstruct a written source for a story or saying that therefore it must derive from oral tradition. In fact it could have been conveyed by any kind of written intermediary now lost to us, from letters to sermons to memoirs to commentaries or histories or anything, not just prior Gospels; and it could also still derive from a prior lost Gospel, or even a Gospel we have, through a completely original retelling rather than a more direct redaction (as evidence suggests John did with the Synoptics: see earlier note 91, page 320), because ancient schools taught *both* methods of composition.[118] The latter fact especially cautions us: it is a grave mistake to assume a redaction must share the same wording and structure as its source material. And finally, such a text could also be a completely original invention of the author (and thus not derive from *any* tradition, oral or otherwise). Or it could indeed derive from an oral tradition—whose content is completely fabricated. All are *prima facie* equally probable. So the task of ruling all these out is not merely daunting, it's often impossible.

I've also seen what I call the Criterion of Repetition: the assumption that just because Jesus is depicted as frequently talking about "The Kingdom," or performing healings, or speaking in parables, it should be concluded that that's what he really did.[119] This is either a non sequitur (such repetition at best only establishes that a particular author wished to *depict* Jesus as emphasizing these things, which then influenced subsequent redactors) or a careless deployment of the Criterion of Multiple Attestation, which we've already seen is of little valid use in Jesus studies (for the reasons there enumerated: see page 172).

We can see a similar criterion in the work of Anthony Le Donne.[120] Le Donne says a great deal of value about how memory becomes distorted (and a historian would do well to heed all he says about that), yet he never presents any valid method for determining whether a claim reflects an actual memory *or a convenient fabrication*. Instead, he relies on the same old criteria proved invalid here (e.g., multiple attestation, coherence, embarrassment) and on classical fallacies like affirming the consequent (e.g., "all apples are red, therefore everything red is an apple"). Le Donne's fundamental thesis, which he employs repeatedly, is that "the more significant a memory is, the more interpreted it will become" and therefore when we find highly interpreted claims in the Gospels he assumes they must reflect a significant memory.[121] For instance he concludes "John was *remembered* as a type of Elijah" (emphasis mine).[122] Yet no valid reason is given for how Le Donne can dismiss the alternative possibility that "John was *represented* as a type of Elijah" for reasons having nothing whatever to do with any actual memory. Instead he just repeats the same fallacies (i.e., either reaffirming the consequent, or appealing to the same invalid "Criteria of Authenticity").[123] Here he applies this 'Criterion of Heavy Interpretation' in conjunction with an invalid EC argument to the effect that "Luke strains to represent Jesus as Elijah, so he wouldn't also repre-

sent John as Elijah"—which is intrinsically illogical. There is no reason
the same type could not be used for both characters, serving different symbolic purposes in each case: in the one instance using Elijah in his role as
the harbinger of the messiah, applied to John, and in the other as a type for
the messiah, applied to Jesus. But Le Donne's reasoning is also invalid for
a more pertinent reason: we know Luke used Mark as a source, so it may
be an association of John with Elijah invented by Mark that had become so
popular (and evangelically useful) that Luke had to use it against his own
literary tendency, and not any actual pre-Markan memory that Luke was
compelled by (or, again, it may have been a pre-Markan invention rather
than a pre-Markan memory). On how we're to tell the difference Le Donne
has nothing to say.

Likewise Le Donne's treatment of Jesus' conflicts with his family, and
every other argument he makes. For instance, he takes the different ways
Mark and John included the "temple/body" resurrection metaphor as evidence of a true historical saying about Jesus destroying and rebuilding the
Jerusalem temple, ignoring the fact that the metaphor appears to originate
with Paul, not Jesus, and thus Mark and John may well be responding to
(or deliberately reconstructing) a fabricated saying, not an actual one.[124]

Paul certainly appears to have originated this "temple/body" metaphor,[125] as well as the "tabernacle made with hands/not made with hands"
distinction as a metaphor for death and resurrection (2 Corinthians 5:1–10),
wherein the one is torn down and the other "built." The saying constructed
in Mark 14:58 (and 15:29; repeated in Matthew 26:61 and 27:40) clearly
alludes to this Pauline teaching because it incorporates "in three days,"
which can only be an allusion to the resurrection of Jesus (1 Corinthians
15:4), and it uses the "made with hands/made without hands" resurrection
distinction exactly parallel to Paul's (2 Corinthians 5:1), and Paul repeatedly equated the body with the temple (1 Corinthians 3:16–17, 6:19–20; 2
Corinthians 6:16). Accordingly, John 2:19–21 simply makes this connection explicit, yet it's already implicit in Mark. Thus Le Donne's interpretation of what Mark is doing with the saying is wholly incorrect.[126] There is
simply no evidence here that Jesus ever really said these things.

Thus Le Donne repeatedly acts like someone who assumes all red things are apples. But just as all red things aren't apples, memories aren't the only things that become highly interpreted (or multiply attested, or mutually coherent, or repeated by later authors contrary to their own literary tendency, etc.). Myths, inventions, and fabrications do as well (just like the many iterations of the Hercules myth). Typology, after all, was more commonly a device used for communicating *ideas*, not memories. Typological constructs in Daniel, for example, do not in any way reflect real memories by or about Daniel. That book is wholly a forgery. How are we to conclude that Jesus is being any more "remembered" in the Gospels than the real Daniel is in the Book of Daniel?

At least Le Donne admits his method gives no certainty, and that there may be no fact of the matter accessible to historians at all (thus he does not even affirm anything about John the Baptist as actually known, not even whether he really was associated with Elijah at any time in his life).[127] But if Le Donne said that of every claim he makes, that would simply be saying we can know nothing about a historical Jesus. Whereas if we agree with his axiom that "a historical argument aims toward the most likely explanation given the historical context and the events that followed by way of impact," then we must obey Bayes's Theorem (the only valid method known for ascertaining "the most likely explanation" among all contenders).[128] And Le Donne simply doesn't.

CONCLUSION

Even the conservative Mark Strauss concludes that Jesus historians have yet to produce any valid methodology from all this confusion of criteria. He observes that these "criteria are often used subjectively and in a circular manner to prove whatever the investigator wishes . . . especially since they can be used to contradict each other," in fact all "too often the criteria are used selectively and arbitrarily to 'prove' whatever the investigator wants to prove."[129] I concur. So does everyone else who's examined the issue. See chapter 1. But any method that makes that possible is clearly invalid and should be abandoned. I've shown that BT replaces *all* the criteria with a valid procedure, and as long as it's used correctly and honestly, it won't let you prove whatever you want, but only what the facts warrant. There is no other contender.

BAYESIAN ANALYSIS OF EMULATION CRITERIA

A completely different set of criteria has been developed by historians of myth and literature that I call "Emulation Criteria" (colloquializing the formal term "mimesis" with the more familiar word "emulation"). The more of these criteria that are met, and the more strongly, the more likely a story in one document is a literary re-crafting of a story in another document. Such literary constructions were not only common in antiquity: the procedure for creating them was specifically taught in schools of the time (again, see note 118, page 323). Can we detect them? Is the procedure of applying Emulation Criteria logically valid?

The best set of Emulation Criteria has been assembled by Dennis MacDonald:

> [There are] six criteria for identifying mimesis [i.e., emulation]: (1) the accessibility and popularity of the proposed model, (2) evidence of analogous imitations of the same story or speech, (3) the volume or number of similarities between two works, (4) the order of the similarities, (5) the presence of distinctive or unusual traits that bind the two works together, and (6) the interpretability of the differences between the two works.[130]

These were inspired by Thomas Brodie, who had earlier formulated three criteria for literary dependence: (1) external plausibility, (2) significant similarities, and (3) intelligibility of the differences (between the emulated and emulating text).[131] Brodie's first criterion MacDonald divides into his criteria (1) and (2); Brodie's second criterion MacDonald divides into his criteria (3), (4), and (5); and Brodie's third criterion becomes MacDonald's criterion (6). Brodie had already articulated his second criterion as keying on "significant similarities" of *theme, pivotal clues, action/plot, unusual details, order, completeness,* and *matching words,* especially *unusual words*, a range of possible parallels that MacDonald simplifies into his three categories: number of correspondences (of whatever kind), order of their arrangement, and remarkably distinctive parallels (such as key words or unusual plot elements).

MacDonald's first and second criteria are not entirely apt. They would presumably establish prior probability: if a particular work or opus (e.g., Homer) was commonly known in the ancient world, the prior probability is increased that that work will be emulated, and if a particular story (e.g., the shipwreck of Odysseus) was frequently emulated in the ancient world (and this second criterion could be expanded to include not just specific stories or speeches, but also frequently emulated characters, like Moses or Odysseus), the prior probability is further increased that that story (or speech or character) will be emulated again. However, in determining the probability of a hypothesis of emulation, the prior probability must actually reflect the frequency with which such emulation is confirmed in a given author and book (e.g., the Gospel of Mark) or in a given type of literature of that general period and place (e.g., ancient hagiography). In other words, if we have caught an author doing this a lot, the odds are high he will have done it in other cases as well; or if it happens a lot in ancient hagiography generally, the odds are high it happened in the Gospels generally. So we should replace his first two criteria with the frequency with which our present author or type of literature deploys literary emulation. To ascertain this without prejudging the conclusion, you could begin with a neutral prior (0.5) and then build a new prior, case after case (using the iteration procedure I described before: see page 168). Nevertheless, if the text or story being proposed as the target of emulation is obscure and unlikely to have been known to the author, we might have to reduce the prior to reflect that fact. But just because the proposed target is ubiquitous, popular, certainly known to the author, and frequently emulated by other authors does not increase the odds that our particular author emulated it, beyond the odds we can already determine that he emulated anything.

So MacDonald's first two criteria even when applicable are only exclusionary—and like all exclusionary criteria, they do not automatically exclude, for the consequents can deviate enough to overcome any prior, no matter how low it is. Thus BT must still be applied to ascertain whether the proposed hypothesis of emulation is probably true (or probably not). The prior we've ascertained for an author or genre (from past cases within that author or genre) is then applied to further cases of suspected emulation in the same author or genre, when the proposed target belongs to the category of popular and commonly emulated books, stories, and characters. But if we suspect an uncommon or rarely mimicked text is being emulated, we should reduce that prior by the degree to which such a specific case of emulation would be unusual (whereas emulating popular and commonly emulated books, stories, and characters is not unusual and therefore requires no reduction of the prior probability). Unusual emulations can thus still be confirmed, if the evidence is strong enough.

BT thus teaches us two things MacDonald's criteria do not: that what matters most is whether we've established a given author is a frequent emulator or emulation is a frequent occurrence in a given genre (which admittedly are the very things MacDonald has set out to establish), and that meeting MacDonald's first two criteria do not increase that frequency (and thus do not increase the odds that the same writer emulated again), but that failing to meet those criteria could decrease that frequency (and thus decrease those odds) in specific cases; yet even that will not entail the emulation did not occur (nor will higher odds entail it did), because we must still evaluate the consequent probabilities. And they can often be overwhelming.

So we turn to MacDonald's other four criteria, which establish consequent probability. First, he looks for "the volume or number of similarities between two works," which (per Brodie) can include similarities of *theme*, *pivotal clues*, *action/plot*, *unusual details*, *completeness*, and *matching words*, especially *unusual words*, and also "the order" in which these similarities appear (which together constitute MacDonald's third and fourth criteria). Many such similarities will often exist merely by chance, or simply because the two stories are about similar topics (like a shipwreck), thus, where h = "emulation occurred," $P(e|\sim h.b)$ will not be low if the similarities are few and already expected. But when the similarities start to accumulate to the point that chance is no longer a credible explanation for why they are there, the scales tip. Such similarities are entirely expected on h, so P(e|h.b) will be high, but such a scale and scope of similarities would be an unusual coincidence otherwise, and unusual coincidences are by definition improbable (unusual = infrequent = improbable), therefore P(e|~h.b) will be low, and by exactly as much as that coincidence is improbable (which obviously will vary from case to case). This is most powerfully demonstrated when very unusual words appear in both stories, or very unusual sequences or events, or anything bizarre, since the inherent probability of such a chance coincidence is always very low, whereas the probability of an emulator borrowing an unusual keyword (or other feature) from the emulated text is always much higher.

An emulating text does not have to have all of these features. It can have any number of them, of any sort. All that matters is that whatever collection of them there happens to be, the odds of all those elements being there by chance (or inevitability) must be significantly less than the odds of their being there by design (and thus as a result of emulation). Because then, the balance of consequents will favor h (often quite strongly). This is further evident in MacDonald's fifth criterion, which calls attention to "the presence of distinctive or unusual traits that bind the two works together," which is just a special case of his criteria three and four: parallels that are distinctive or unusual in both stories (and thus not commonly found in any other stories, even of the same type). These are elements that are already improbable in and of themselves, so to see them in both stories *alone* often swings the balance of consequents in favor of emulation.

As with the trustworthy neighbor example in chapter 3 (page 74), the *actual* consequent probabilities here could both be extremely low (since we usually can't predict from h exactly which words or ideas will be used or that any specifically will), but all that matters is their ratio, and using any particular keyword or concept from an emulated story is always more likely on h than chance (where h = emulation and ~h = chance), so when enough of these parallels are present, the effect can be huge. So we can treat the consequent for h as being effectively 1 and setting the consequent for ~h in ratio to that (I discuss this practice of disregarding contingencies in chapters 3 and 6: see pages 77–79 and 214–18).

Nevertheless, when we do this, "chance" does not then mean the bare probability of assembling a particular collection of words and ideas in one place. You must avoid the Lottery Fallacy, the idea that because winning is improbable therefore the player must have cheated—when in fact, odds are, *someone* was going to win, because there are so many players (this fallacy is discussed in chapter 6: see page 227). Related to this is the Fallacy of Multiple Comparisons, which must also be avoided when looking for emulation: if we allow any comparison to any text, odds are we'll always find some similarities simply by chance—this simply won't be unlikely at all. There must be some constraints on what counts as a parallel and what doesn't (which is the function of MacDonald's criteria), and there must be too many parallels for multiple comparisons to have caused them. Accordingly, because there are so many texts and so many ways to say the same thing, when all the coefficients of contingency cancel out, the consequent probability for ~h can often end up as near to 100 percent as the consequent for h, no matter how unique the text may be or what coincidences appear in it. Because as discussed in chapter 3 (pages 77–78), every configuration of words is extremely improbable, so what matters is their *generic* likelihood, which can often be quite high.

Thus the coincidences that you propose are emulative must be truly unusual, highly numerous, and/or unarguably apposite in meaning, none of which chance can easily explain. In other words, you must attend to the difference between (a) calculating the odds of finding someone who won a lottery, which can be near 100 percent no matter how unlikely winning that lottery is (hence when someone wins a lottery we usually infer chance, not design), and (b) calculating the odds of accidentally banging out an entire sheet of Chopin's music by randomly hitting piano keys, a case where the specificity is huge in relation to the number of available attempts, no matter how many attempts are made (hence when some stranger rattles off a bit of Chopin we infer design, not chance). Like finding a lottery winner, P(e|~h.b) must reflect the probability of a chance coincidence *all else considered.* Hence, it can take a lot to lower it. Basically, the question you must ask is, can we reasonably expect to find any set of coincidences like that anywhere in ancient literature just by chance? If the answer is yes, then P(e|~h.b) is high. If no, then it's low.

Such follows for evaluating the similarities. What about the differences? Contrary to common assumption, the differences between the two stories will not lower P(e|h.b). As MacDonald rightly argues, some differences actually argue *for* emulation, and few will ever argue against it, because emulation *entails* the implementation of differences. An author only wants to adapt select elements of an emulated story to develop an otherwise entirely different story, and the emulating author will always have different literary interests than the emulated author had—in fact, as MacDonald aptly explains, often the emulator's interests are to *subvert* the emulated tale (as when Virgil subverts the stories in Homer in order to demonstrate how Roman beliefs and values are superior to Greek). Hence, sometimes the differences are the whole point of the emulation, allowing an attentive or clued-in reader to extract the intended meaning from precisely what has changed. We can detect this when differences from the first story make unusual elements of the second story more intelligible on *h* (criterion 6), or when differences that exactly reverse the order or gender or other element (criterion 4) make *h* more (or at least as) probable as ~*h*. And when differences are already contextually expected on *h* (like changing the geographical location or the exact metaphor used to symbolize the same story element), this won't make *h* less probable; nor will any other differences that serve the author's interests and aims.

Indeed, differences that increase interpretability actually increase P(e|h.b), because that's exactly what we expect on *h*, and they also *decrease* P(e|~h.b), because it would be unusual if a purely chance inversion of the first story made the second story more intelligible. Such a coincidence is not impossible, just improbable, and that improbability must still lower P(e|~h.b), and by exactly as much as that correspondence is improbable (at least on any other cause than authorial design). This is MacDonald's sixth criterion ("the interpretability of the *differences* between the two works"). But that criterion should be expanded to include *any* increased interpretability, even if it derives from a similarity and not a difference (a fact perhaps meant to be captured by MacDonald's fifth criterion).

Differences that involve an exact reversal of elements also won't usually reduce P(e|h.b) because such reversals frequently occur in emulation. But sometimes exact reversals will lower P(e|~h.b), e.g., using the exact five words in reverse still entails an unexpected coincidence on ~h, just as would using the exact five words in the same order. Random chance can produce many three- and some four-word exact sequence matches, so the amount these lower the consequent on ~h will likely be so small as to be washed out by *a fortiori* estimates (unless someone can perform a statistical analysis on the Greek corpus showing otherwise). Likewise single-word match-ups will be common by chance, except in the case of very rare words, or words that are very unexpected otherwise, or when there are so many one-word match-ups as to be demonstrably unusual.

All the same goes for plot elements such as the sequence of events, or switching genders or other categories (e.g., from indoors to outdoors or from war to peace). Again, such reversals might not argue *for* emulation, but neither do they usually argue against it. But if such exact reversals are entirely expected on *h* (i.e., if *h* makes exactly that reversal very likely), then they *do* argue for emulation: they will increase P(e|h.b), because such an observation is expected on *h*, or else (or also) decrease P(e|~h.b), since a reversal that's expected and intelligible on *h* is often also an unexpected coincidence on a hypothesis of chance, and thus less probable on ~h.

MacDonald's works are filled with examples of applying these criteria. If we structured his arguments to conform to BT, I think we'd find that most of those examples would produce only a weak conclusion (like a maximum P(h|e.b) of around 60 percent), but many will clearly entail a strong conclusion (like a P(h|e.b) greater than 90 percent or even 99 percent or well above). Doing this would not require any exact knowledge of any of the relevant statistics, since easily adduced *a fortiori* estimates will usually suffice (and if anyone suspects otherwise, they can always generate the relevant statistical data to find out). MacDonald is also not alone. Thomas Brodie has surveyed many other examples; and Randel Helms has added many more (though without explicit application of criteria). The same BT analysis can test their claims as well. I will demonstrate many of the best examples (my own and theirs, and those discovered by other scholars as well) in my next volume. Here I will provide just one example (following) to illustrate the concept. But as just shown, the method they are all using is valid and sound (when competently employed), and can be verified as such with BT: when the overall features of a story are significantly less probable by chance than by emulation, emulation has probably occurred, especially if we can show that the same author does this a lot. The significance of this for historicity is that a story that was probably produced by emulation is less probably a historical fact; and an author who composes primarily by emulation probably isn't writing a history. But I won't demonstrate those inferences here.

"Daniel in the lion's den" becomes "Jesus in the empty tomb"

To illustrate the above I will draw on an example I published myself.[132] We know from archaeology that the story of Daniel in the lion's den was a popular symbol of resurrection (and of Jesus) among early Christians.[133] The story is told in the Old Testament book of Daniel, which we know was written (and translated into Greek) over a century before the time of Christ, and was very popular. As the story goes, when Daniel was entombed with the lions, and thus facing certain death, the Persian king Darius placed a "seal" on the stone "so that nothing might be changed in regard to Daniel" (Daniel 6:17). The same thing is done in Matthew's story of Jesus' burial: the Jewish authorities place a "seal" on his tomb, and post a guard, so they could be sure his body stayed put—a whole incident that conspicuously doesn't occur in any other Gospel, not even Matthew's source, Mark, whose account (as also Luke's and John's) thoroughly contradicts Matthew's additions in this regard (thus confirming they are fictional creations of Matthew—see previous discussion in this chapter, page 128).[134] The absence in every other Gospel, especially Matthew's source Mark, of guards, seal, even prior awareness of a possible plan to steal the body, as well as an ensuing miraculous angelic act, is less likely on the theory that any of this actually happened, than on the theory that Matthew made it all up, intentionally embellishing the story he received from Mark. This fact alone entails P(e|HISTORICAL.b) $\ll$ P(e|MYTHICAL.b). But we needn't stop there.

In Matthew the placing of the seal (Matthew 27:66) is described with the exact same verb used in the Greek edition of the Daniel story (*sphra-gizô*), which is in both stories a rather unusual detail. This evokes a meaningful parallel: Jesus, facing real death, and sealed in the den like Daniel, would, like Daniel, escape death by divine miracle, defying the seals of man. The parallels are too dense to be accidental: like the women who visit the tomb of Jesus at the break of dawn (Matthew 28:1), the king visits the tomb of Daniel at the break of dawn (Daniel 6:19); the escape of Jesus signified eternal life, and Daniel at the same dramatic moment wished the king eternal life (Daniel 6:21; cf. 6:26); in both stories, an angel performs the key miracle (Matthew 28:2, Daniel 6:22); and after this miracle in Matthew, the guards curiously become "like dead men" (Matthew 27:4) just as Daniel's accusers are thrown to the lions and killed (Daniel 6:24). The very unusual choice of phrase "like dead men" in Matthew thus becomes explicable as an allusion to these victims in Daniel. The angel's description is also a clue to the Danielic parallel: in the Septuagint version of Daniel 7:14, an angel is described as "and his garment white as snow"; in Matthew 28:3, the angel is described as "and his garment white as snow," in the Greek every word identical but one (and that a cognate), and every word but one in the same order. Another angel in Daniel 10:6 is described as "his outward appearance as a vision of lightning" while the angel in Matthew is similarly described as "his appearance as lightning." The imagery is thus a Danielic marker: Matthew is getting his ideas of what an angel looks like and how to describe one not from eyewitnesses *but from Daniel*, exactly where he's getting the lion's den story. (Matthew was fond of expanding on Mark by lifting ideas from Daniel, e.g., Matthew 17:6–8 takes material from Daniel 10:7–12 to expand on elements of Mark 9.)

Furthermore, Matthew alone among the Gospels ends his story with a particular commission from Jesus (Matthew 28:18–20) that matches many details of the ending of the Greek version of Daniel's adventure in the den: Jesus says God's power extends "in heaven and on earth," to "go and make disciples of all nations" and teach them to observe the Lord's commands, for Jesus is with them "always" even "unto the end." And so King Darius, after the miraculous rescue of Daniel, sends forth a decree "to all nations" commanding reverence for the Jewish God, who lives and reigns "always" even "unto the end," with power "in heaven and on earth" (Daniel 6:25–28). The latter phrase in Greek is even identical in both cases. The stories thus have nearly identical endings. Indeed, the king's decree in Daniel reads like a model for the very Gospel message itself (see Daniel 6:25–27). And the episodes are framed the same way: in both Matthew and the Greek text of Daniel the stories introduce their parallel structure with the same verb and object, "to seal" (*sphragizô*) the "stone" (*lithon*), and conclude it with the same teaching about the Lord reigning until the "end" (*telos*; *sunteleia*) of the "eons" (*aiôn; aiônas*).

Since the placing of a "seal" is essential to creating the Danielic parallel, Matthew has a motive for inventing the entire motif of the guards in order to create the pretext, not only for the sealing, but for the clue of "becoming like dead men" and the angelic "miracle," all elements unique to his story. There are more telltale signs that this story is fabricated, and that Matthew fabricates other stories like this with some frequency, but what I've summarized here is enough for the present purpose. The evidence e thus includes not only the fact that this entire story is unexpectedly unique to Matthew (despite our having three other versions to compare, including Matthew's own source), but also all those other facts that link his unique changes to the story to the book of Daniel and its tale of the lion's den. Since the latter is entirely expected on the hypothesis of fabrication, but not as expected on the hypothesis of historicity, what was already $P(e|\text{HISTORICAL}.b) \ll P(e|\text{MYTHICAL}.b)$ becomes $P(e|\text{HISTORICAL}.b) \lll P(e|\text{MYTHICAL}.b)$. We must also begin with $P(\text{HISTORICAL}|b) \ll P(\text{MYTHICAL}|b)$ owing to the fact that integral to the story is a grandiosely flying, paralyzing angel, which is inherently more likely to be made up than true (since we just don't see such things happening in the real world—thus, if they happen at all, it's but rarely, whereas made-up stories about magical beings are extremely common)—the more so if we can establish a trend of fabrication and mythmaking in Matthew (and we can), but we needn't add that in here. Just with what we have enumerated here, no matter what numbers are entered in, the end result is going to be $P(\text{HISTORICAL}|e.b) \to 0$.

Focusing on the emulation hypothesis alone, all six of MacDonald's criteria for literary emulation are met: the text being imitated (the

Septuagint, which by then included the book of Daniel) was well-known and frequently used this way, and the comparison of Jesus with Daniel was a common one (so there are no deductions from the prior probability on that account); there are several significant parallels; the parallels often appear in the same order; the connection is confirmed by peculiar features (direct borrowing of terms and phrases; the unusual description of the guards); and the whole device reveals an obvious, intelligible meaning. Indeed, the story becomes interpretable, with obscure and seemingly confusing features suddenly making perfect sense (such as why the guards become "like dead men," why the Jews bother with a seal when they have a guard, why an angel has to intervene even though Jesus is apparently no longer even in the tomb, and why Matthew alone reports any of this). All of these factors are more expected (and thus more likely) on the hypothesis of emulatory fabrication than on any other hypothesis. For all these coincidences with the Daniel story to have *actually* happened is intrinsically improbable (even if not impossible, certainly not highly probable—in fact it's highly *improbable*, the likes of which never found by chance anywhere else), but for these coincidences to exist as a product of emulatory fabrication is intrinsically likely (in fact, entirely expected).

Illustrating the principles set forth earlier, the fact that the description of the angel lifts a whole string of words directly from Daniel is very likely if Matthew is getting it from Daniel, but extremely unlikely if he is not. The probability of such a coincidence of words and word order being a product of random chance is extraordinarily small. The fact that one word is switched for a cognate, and one word is in reverse order, makes no significant difference to this calculation. One can propose the alternative theory that Matthew borrowed a description from Daniel to embellish an otherwise true account, but then you have the improbable coincidence that Matthew's story uniquely draws ideas from and makes allusions to a story unique to the book of Daniel, indeed one in which an angel also performs the key miracle in the narrative, the very angel Matthew then endeavors to describe by lifting angelic descriptions from that very same book. This coincidence is expected on the hypothesis of fabrication, but much less expected on the hypothesis that he is only embellishing a true story.

Likewise, all the deviations do not alter this ratio. For example, the fact that there are no lions in Matthew's version of the story does not make borrowing less likely, since removing them is exactly what we would expect (since Matthew is starting with a narrative from Mark that already excludes them, and in which introducing them would make no sense). The fact that "the chief priests and the Pharisees" place the seal, instead of a king (much less King Darius), does not make borrowing less likely, since Matthew's story already begins in a context that makes that particular switch likely: he is setting his own story in a different historical and political context, and has different identifiable literary aims, which both entail that elements of the Daniel story unsuited to his expected context and purpose will be dropped or altered (and no emulation hypothesis entails *exactly* which elements will be adapted, only that more will be than chance alone can easily explain).[135]

In short, none of the changes make emulation less likely. But many elements do make it more likely; and some *changes* even make emulation more likely. For example, unlike in the lion's den, the guards are only paralyzed, yet oddly said to "become like dead men," rather than actually being killed (much less by the ruling parties). But these differences are expected. The motive and occasion for killing Daniel's accusers doesn't exist in Matthew's story, and killing the guards would destroy Matthew's intended plot, which requires the guards to lie about what happened, and his story wouldn't work at all if the Christians were on the hook for murder. But that the guards are only described as "like dead men" is actually itself an improbable coincidence on the theory that Matthew *isn't* alluding to (and thus borrowing the idea from) the lion's den tale, thus though marking a *difference* between the two stories, this actually *reduces* $P(e|\text{HISTORICAL}.b)$ and *increases* $P(e|\text{MYTHICAL}.b)$. This is because such a description is strange and unexpected—*unless* it's an allusion to the lion's den tale, in which case it's neither strange nor unexpected (or certainly much less so). Overall, that chance coincidences of history or storytelling would produce *all* these congruences is very improbable, but that emulating Daniel's tale of the lion's den would produce them all is far more probable. Thus the consequents favor emulation—especially when combined with all the other evidence (such as the contradictory silence of other Gospels).

This is just one example, and indeed it's not even the best. I chose it because of those features allowing us to confirm the story really is fabricated *independently* of any application of emulation criteria. But even using those criteria alone it's sufficiently strong to be clear (to anyone not dogmatically set against the conclusion) that Matthew made all this up to equate the tomb of Jesus with Daniel's den of lions. MacDonald's criteria pertain, confirming that what is evident is genuinely there. And their capacity to do this is entirely explained and validated by Bayes's Theorem.

BAYESIAN DEMONSTRATIONS OF AHISTORICITY

In the last three chapters I've made a strong case that all valid historical methods are described by BT and that BT proves the methods so far used by Jesus historians are either invalid or invalidly employed. It follows that we must use BT.

Applying BT to the specific question of whether some person, place, or event actually existed or not is merely a matter of ascertaining the prior and consequent probabilities. What is the prior probability that Jesus actually existed? What is the prior probability that he didn't? How likely is the evidence we have if he did exist? How likely is the evidence we have if he didn't? Given the most obvious answers to these questions, at first glance it seems surely "Jesus existed" would win out as the most probable hypothesis on BT. In my next volume (*On the Historicity of Jesus Christ*) I'll reveal that on *second* glance, that conclusion is not so obvious, and might even be wrong. But you needn't believe that now. I'm obligated to prove it, and to persuade my expert peers thereby. If I can't, I'm wrong. But I won't undertake that task here. Here I will only close with the required methodology (adding to the relevant remarks on this task already concluding chapter 4).

The two hypotheses to test will be h = "Jesus was a historical person mythicized" and $\sim h$ = "Jesus was a mythical person historicized." The only other logical possibilities are h_0 = "historical person not mythicized" and $\sim h_0$ = "mythical person not historicized," but our background evidence firmly establishes the prior probability of either of those is vanishingly small (all reasonable Jesus scholars agree Jesus was mythicized to some degree; and even those who would deny he existed agree he was historicized), while the consequent probability of the evidence favors neither of those over h and $\sim h$ (i.e., h and $\sim h$ make all the evidence just as likely or far more so than either h_0 or $\sim h_0$). So we can disregard h_0 and $\sim h_0$ (their consequents lend them no credence and their priors are so small they won't even be visible in our math). So the prior probabilities of h and $\sim h$ must still in effect sum to one. I will present a method for determining their value in *On the Historicity of Jesus Christ*, after more carefully defining h and $\sim h$. That leaves estimating their consequents, which will then occupy the rest of that volume.

If it's more inherently likely that a savior god like Jesus would be mythical, then the priors will favor $\sim h$, and if proposing this actually explains all the evidence better than any alternative, then the consequents will also favor $\sim h$. And if both the priors and consequents favor $\sim h$, then we must conclude that Jesus probably didn't exist. But all these values are conditional on background knowledge. And that's where the debate often focuses, and contemporary scholars feel certain that deniers of the historical existence of Jesus fail to grasp the extent and significance of that background knowledge, and thus grossly misestimate the probabilities of different theories of the evidence. Deniers, meanwhile, charge historicists with ignoring telltale oddities in the evidence that make little sense unless Jesus never really existed to begin with and was only created after the fact. Who is right deserves another look. But that's for another time. What has been shown here is that it is at least logically possible to prove from existing evidence that Jesus (probably) didn't really exist, and how one can legitimately do that. Whether the evidence goes that way, however, cannot be presumed.

CHAPTER 6
THE HARD STUFF

No presently articulated system of formal logic is really very relevant to the work historians do. The probable explanation is not that historical thought is nonlogical or illogical or sublogical or antilogical, but rather, I think, that it conforms in a tacit way to a formal logic which good historians sense but cannot see. Some day somebody will discover it, and when that happens, history and formal logic will be reconciled by a process of mutual refinement.

— David Hackett Fischer[1]

The preceding chapters established my case for adopting Bayes's Theorem as the standard model for reasoning among historians, and explained most of the basics of how to go about doing that. With the many examples provided, the preceding ideas and information can be adapted to all other cases and circumstances. This final chapter will change tack and address deeper issues regarding the application and applicability of Bayes's Theorem generally.

Six issues will be taken up here: a bit more on how to resolve expert disagreements with BT; an explanation of why BT still works when hypotheses are allowed to make generic rather than exact predictions; the technical question of determining a reference class for assigning prior probabilities in BT; a discussion of the need to attenuate probability estimates to the outcome of hypothetical models (or a hypothetically infinite series of runs), rather than deriving estimates solely from actual data sets (and how we can do either); and a resolution of the epistemological debate between so-called 'Bayesians' and 'frequentists,' where I'll show that since all Bayesians are in fact actually frequentists, there is no reason for frequentists not to be Bayesians as well. That last may strike those familiar with that debate as rather cheeky. But I doubt you'll be so skeptical after having read what I have to say on the matter. That discussion will end with a resolution of a sixth and final issue: a demonstration of the actual relationship between physical and epistemic probabilities, showing how the latter always derive from (and approximate) the former.

RESOLVING DISAGREEMENTS

In chapter 3 I already argued that BT does not create any more disagreements over probabilities than any other method entails—rather, it exposes to the light of day disagreements that already exist and should be resolved anyway. I won't repeat that argument here, but rather expand on one element of it: the question of how such disagreements can be resolved.

Bayes's Theorem should first be used to calculate what you yourself should believe given what you can honestly claim to know. This amounts to a process of checking your belief system for logical inconsistencies and correcting them. But eventually you will want to persuade others that your conclusions are correct. This requires that three basic conditions be met: anyone you intend to persuade this way must be committed to being reasonable and objective; they must accept the validity of Bayes's Theorem and understand its mechanics; and they must share the same relevant expert knowledge. The first condition is fundamental—anyone who fails to meet it should simply be ignored no matter what the case, since the opinions of such people are of no interest to serious scholarship. The second condition can be taught (using the resources in this book and those referenced in chapter 3). And the third condition can be realized by the exchange of information (including primary evidence and published scholarship). But even when these conditions are met there will still be disagreements. What follows is a logical procedure for resolving those disagreements when those three basic conditions have already been met. Some of this overlaps what was said about this in chapter 3 (page 88), where I already argued that adequate communication will likely resolve most disagreements by establishing their irrelevance or increasing agreement by increasing available information (which results in opinions converging on a common *a fortiori* answer). Here I will add more technical advice.[2]

The most common disagreements are disagreements as to the contents of *b* (background knowledge) or its analysis (the derivation of estimated frequencies). Knowledge of the validity and mechanics of Bayes's Theorem, and of all the relevant evidence and scholarship, must of course be a component of *b* (hence the need for meeting those conditions before proceeding). This basic process of education can involve making a Bayesian argument, allowing opponents to critique it (by giving reasons for rejecting its conclusion), then resolving that critique, then iterating that process until they have no remaining objections (at which time they will realize and understand the validity and operation of Bayes's Theorem and the soundness of its application in the present case). So, too, for any other relevant knowledge—although they may also have their own information to impart to you, which might in fact change your estimates and results, but either way disagreements are thereby resolved as both parties become equally informed and negotiate a Bayesian calculation whose premises (the four basic probability numbers) neither can object to, and therefore whose conclusion both must accept.

In complex cases, arriving at such an epistemic equilibrium requires continual and persistent dialogue such as asking each other questions to determine what actually differs between you in respect to the content of *b* or its analysis and why. This also follows for disagreements regarding the contents of *e* (the evidence to be explained by the hypothesis) or the exact formulation of *h* (the hypothesis) or its competitors (all other hypotheses subsumed within ~*h*). Only once agreement is reached on these matters will the results of Bayes's Theorem be the same for all of you. Such a process should lead to each of you purging elements (of *b*, *e*, or *h*) that are not in fact defensible or appropriate (e.g., assumptions you cannot demonstrate are correct, claims you cannot establish, etc.) and/or accepting new elements (of *b*, *e*, or *h*) that are defensible or appropriate (e.g., new assumptions that have been demonstrated are correct, new claims that have been established as true, etc.). Eventually, through such a negotiation, you will mutually agree on the acceptable contents of b, e, and h (and any analysis therefrom), and from this will necessarily follow the same conclusion for all of you via Bayes's Theorem.

There will remain occasions when you will have access to information the other can never access (usually private unshared experiences), in which case you will each get a different result from Bayes's Theorem. But since BT only produces a conditional probability (it demonstrates what your conclusion should be *given what you know*), disagreements in this case will be acceptable to both parties. Once all other disagreements are resolved in the manner described above, Party A will agree that Party B's conclusion should in fact be exactly what B finds it to be, given the information available to B, and Party B will agree that Party A's conclusion should in fact be exactly what A finds it to be, given the information available to A. In other words, they will actually agree they must disagree, and in exactly the way determined by their different results with Bayes's Theorem, precisely because they each have access to information the other cannot confirm. Each will thus agree the other's position is entirely rational (provided they've been sincere and are not insane), and therefore their disagreement is entirely appropriate. This latter condition does not support claims of epistemic relativism, however, since there is still a single objective fact of the matter (one or both parties are still wrong); it's just that to one (or both) of them the required information is unavailable and we must all work from what we know. The films *Contact* and (the original) *Journey to the Center of the Earth* each present clear (though far-fetched) examples of entirely valid instances of just such a condition, where one party validly knows the truth but cannot expect anyone else to agree with them. And in such cases the appropriate attitude of everyone else should be the same: that the party making the claim cannot be expected to disavow their conclusions, provided they in turn accept that others cannot be expected to share those conclusions.

However, this does not give warrant to every personal belief, as there must still be agreement on all other mutually accessible facts and their analysis. For example, if Party A has visions of deity B, it does not follow that they have personal unshared knowledge of that deity (or any deity at all), since *b* (for anyone informed as they should be) must include knowledge of the cultural, psychological, and biological causes of such experiences and their documented effects (worldwide and throughout history), which are highly various and mutually contradictory. As this is knowledge accessible to everyone, including Party A, it entails Party A should be skeptical of their visions (just as they would be skeptical of another party claiming visions of an entirely different deity).[3] I have provided examples of this from my own experience, in each case rejecting the *prima facie* implications of my direct personal experience in consequence of my scientific background knowledge.[4]

On the other hand, it is always possible (and in fact must be the case and is very routinely the case) that trust can be sufficient to warrant accepting data you cannot personally access but that others attest to. In such cases a separate Bayesian analysis would show that in those cases you should trust what is reported and include it in your own *b* or *e*. This conclusion only fails to follow when a Bayesian analysis determines that such trust is unwarranted or must be attenuated to some nontrivial probability (as the probability of its being wrong is no longer vanishingly small). At least in the latter case you can sometimes arrive at a conclusion that what they report is probably true (and a sound Bayesian analysis will determine this for you). But that becomes a hypothesis to test, that is, it's not a given, but a conclusion you have to argue for. To treat it as an established fact usually requires more. Of course most "publicly available data" will consist of testimonies to facts not independently accessible to the historian, but in this case all historians are in the same relationship to the evidence (i.e., the data that *is* available is equally accessible to all). Cases of rationally warranted disagreement arise only when one historian actually has access to data that other historians do not—and then only when their testimony to that data is insufficient to be universally trusted even when granting its sincerity (e.g., the Tibetan peasants seeing a giant Buddha in the sky, as analyzed in chapter 3, page 72), or precisely because its sincerity can't be granted (e.g., scholars often don't have adequate warrant to be so trusting when a scholar claims to have consulted a source or to have seen evidence that she can no longer produce).

Hence the reason public or replicable data is so important to professional history (per my first axiom in chapter 2, page 20) is that it allows us to personally observe the same data (thus bypassing the need to trust more people than we have to), and the reason expert consensus is so important (per my second axiom in chapter 2, page 21) is that when the competent reporting witnesses are extremely numerous (e.g., a whole community with considerable training, mutually policing effective standards), the probability of mass error or deceptive collusion becomes extremely small. I'll revisit this point briefly later, and I discussed a few examples in chapters 2 and 3, but I won't analyze when and why to trust experts here. That will already become part of the information-sharing dialogue between disagreeing parties. And such dialogue almost invariably creates agreement. Rationally justified disagreement among well-informed parties is comparatively rare.

Setting aside those cases of rational disagreement, what remains is professional agreement, that is, agreement on what historians as a group should declare to be known. In other words, conclusions on which we should rationally expect all historians to agree, because none of the determining data is inaccessible to them. Resolving such disagreement begins with achieving agreement on the contents of e and then b, which requires isolating what those contents are (as far as pertains to the disagreement), which must include agreement on what is *not* in e or b.[5] This should be resolved first. If disagreement persists even after agreeing on that, the debate next moves to achieving agreement on the content of h and its considered competitors, which will naturally merge with the next step after that, achieving agreement on what h entails as regards predicted effects (i.e., what evidence is likely or unlikely given h or its competitors). Once agreement is also reached on both of those points, for every viable hypothesis, all that remains is to agree on priors and consequents. As noted in chapter 3 (page 89), strict agreement is unnecessary here. Only when disagreements on priors or consequents actually entail different conclusions (and that means in the more general sense of only ruling a claim "true," "false," "likely," or "unlikely") do those disagreements matter.

Resolving such disagreement requires exploring why each party derives the probability they do, and why they differ despite deriving it from the exact same information. One can do this by identifying determinable probabilities that can be connected to the probabilities being estimated and ask why deviations obtain. For example, if two historians disagree on how frequently bodies were stolen from graves in antiquity, at least one undeniable limit can be established: a maximum number of bodies available to be stolen in a given year can be agreed upon (which, let's say, archaeology can confirm can't have been more than 1,000,000 for any particular graveyard), and if both parties also agree at least one of those bodies would be stolen each year, you have a definite minimum frequency (one in a million per year), and if one party's estimate was lower than that, they must now agree to revise it. Thus at least some kind of minimum can be arrived at. Then you can approach the matter from the other side: if one party insists the frequency cannot be as high as, say, 1 out of every 1,000 bodies, yet this is their opponent's *a fortiori* maximum, their opponent must ask *why* they conclude the rate can't have been that high. If they can give no valid reason, then their objection is without foundation—they must then agree the rate could have been that high, as they know of no valid evidence it was lower. Now with a working maximum and minimum, a calculation can be made. And sometimes only the maximum matters to the actual argument being made, for example, if it is argued "so far as you know 1 in 1,000 bodies were stolen in any given year," then any conclusion that follows from this can also be argued to hold "so far as you know" (because any qualifiers in the premises will commute to the conclusion). Thus the conclusion "so far as you know P(h|e.b) = x" would have to be accepted by both parties (otherwise one of them is rejecting sound logic). In this dialogue, all relevant evidence could be adduced regarding the frequency of bodysnatching (e.g., from laws passed, cases recorded, etc.) and similarly debated in respect to the minimum and maximum rates that would explain all that evidence, and if disagreements persist even there, the same debate can surround why. Finally, both parties can discuss what further inquiry (by collecting more information) might change their minds—and if that inquiry is possible, the prescribed research can be completed and the issue revisited.

This example may seem silly, but the principles it exemplifies can be adapted to any substantial dispute over probabilities.[6] It's still worth repeating that most such disputes won't matter and thus needn't occupy anyone's time. If the differing estimates of each party both produce essentially the same conclusion, then you have all the agreement you need. No further discussion is necessary. There is also a middle ground historians can explore: rather than insisting a particular model is correct, instead build a model to ascertain what assumptions are necessary for that conclusion to obtain. Carrying forward the same example above, BT can be used to demonstrate that a particular conclusion requires a particular minimum or maximum frequency of body snatching, which knowledge can be useful in and of itself without requiring commitment to these frequencies or anything that follows from them. A similar approach can be used to determine if further inquiry on a particular question will be fruitful, to determine what further evidence you should be looking for (to verify or falsify a preliminary result), or just to see what possible scenarios can fit the evidence and how credible they thereby become. Since, as noted in chapter 3 (page 61), *merely* fitting the evidence does not make a theory credible at all. So BT can be deployed to ascertain if such a fit has that result or not.

THE VALIDITY OF NONSPECIFIC PREDICTION

In chapter 3 I argued that it's not necessary for a hypothesis being tested in BT to make exact predictions of what evidence will exist (beginning on page 77). It's sufficient to construct h to make only generic predictions (predictions of what type of evidence to expect). Although of course h could be constructed to combine both types of predictions. And in science h often at least seems strictly constructed to make only exact predictions. But the latter is illusory, as I showed by pointing out that even scientific hypotheses, no matter how strictly constructed, still ignore all manner of details such as exactly which scientist will make the predicted observations or at exactly what time of day, and this same conclusion was all the more obvious in historical sciences like geology (see chapter 3, page 46). Thus, in practice, *all* applications of BT, even in the hardest sciences, make predictions from *h* regarding the contents of *e* that are to some extent generic and not exact.

I've encountered even mathematicians who react to this with suspicion, though I don't understand why. The mathematical justification should be obvious to them if anyone.[7] Consider the configuration of the stars in the sky: the probability that the stars would today stand in exactly the pattern they now do is vanishingly small, whereas an intelligent engineer who intended to put them in exactly that pattern would make that pattern 100 percent certain. But surely that does not mean that the stars must be in that pattern because of intelligent design. From the conditions fixed shortly after the Big Bang, that the stars would exist now in *some* pattern that is similarly complex and comparably arranged is all but 100 percent certain, so their existing in that pattern is no more likely on design than natural causes. This is because the stars would inevitably come out in some such complex and unique pattern. So appealing to the complexity of the pattern is fallacious, since in all probability no matter what pattern it came out to be, it would have been just as complex, yet *in the same generic features entirely the same.*[8] Thus, the Big Bang Theory does not predict exactly how the stars would be arranged; it predicts only what general pattern they would exhibit (which pattern can be defined by a mathematical formula that would describe *all* equally likely arrangements, and that would entail conclusions like "they will not likely form a perfect cross in the center of the sky as viewed from Jerusalem at midnight every Yom Kippur"). There are thus countless ways the stars could have been arranged that would verify the theory, and as long as they are arranged in any of those ways, we can declare $P(e|h.b) \approx 1$ (where e = the configuration observed and h = the Big Bang Theory).[9] But someone might still object to this conclusion. Hence the following discussion.[10]

I'll start with a different example expanded from chapter 3. A Bayesian analysis of a drug's efficacy will ignore such contingencies as the name of the scientist who will observe and report the results. But technically we might object to that, since "result x will be reported by Dr. Smith" and "result x will be reported by Dr. Jones" are two different outcomes, and thus in each case we have a different e, one in which we have data from Dr. Smith, and another in which we have data from Dr. Jones. We therefore cannot say e is, for example, 100 percent likely if h is true (i.e., that h strongly predicts e) when we could have had a different e, i.e., we could have data from Smith instead of Jones. If the hypothesis is that the drug will always show outcome x, we obviously want to say $P(e|h.b) = 1$ when $e = x$, but that's impossible when 'x from Smith' and 'x from Jones' are mutually exclusive, yet both outcomes are equally possible on h (just as countless different configurations of the stars are possible on the Big Bang Theory). Since e is always one or the other (i.e., either from Smith or from Jones), and nothing in h entails one over the other (i.e., neither scientist's being the observer is more likely), at the very least $P(e|h.b)$ should be 0.5 for each possible outcome (either an observation of x by Smith or an observation of x by Jones). But the number of possible scientists is factually in the thousands, and in hypothetical extension approaches infinity (i.e., there are infinitely many "Dr. Z saw x" outcomes that are logically possible yet would still fulfill the prediction of h, and as shown in chapter 2, page 23, nearly everything that is logically possible has a nonzero probability).[11] So, too, the configuration of the stars.

It seems intuitively obvious that this is ridiculous and that we're right to ignore these contingencies, but on close reflection it's not immediately clear why that intuition is correct (which I suppose explains those mathematicians who scoff when I suggest it). The justification is fairly simple, however. Since h actually makes no predictions regarding who will make the observations and who won't, the coefficient of contingency will be the same for both consequent probabilities. For example, assume the probability is n that x will be observed by Smith rather than any other of the thousands of scientists who realistically could have been in her position, and that the probability that x will be observed if h is true is otherwise 1 and the probability that x will be observed if h is false is otherwise (let's say) 0.2, and that the priors are equal. Then on some scientist observing x the completed formula would be:

$$P(h|e.b) = \frac{P(h|b) \times P(e|h.b)}{[\ P(h|b) \times P(e|h.b)\ ] + [\ P(\sim h|b) \times P(e|\sim h.b)\ ]}$$

$$P(h|e.b) = \frac{0.5 \times (1 \times n)}{[\ 0.5 \times (1 \times n)\ ] + [\ 0.5 \times (0.2 \times n)\ ]}$$

$$P(h|e.b) = \frac{0.5n}{0.5n + 0.1n} = \frac{0.5n}{(0.5 + 0.1)n} = \frac{0.5}{(0.5 + 0.1)}$$

The coefficient n thus cancels out. It vanishes from the final equation. It therefore never needed to be introduced in the first place. This is because any probability that, say, Smith is more likely to have observed e is the same whether h or $\sim h$ and therefore "the probability that e would be observed if $\sim h$" is multiplied by exactly the same probability (that it would be Smith instead of Jones who saw e) as "the probability that e would be observed if h." Since it's just as likely that Smith would be the one to observe e whether h or $\sim h$, it doesn't matter what the likelihood is that Smith is the one to have observed e (and, as here shown, Bayes's Theorem proves this). Just as this is true for the exact name of the scientist who makes the observations that become e, it's true of every other contingency of whatever kind (such as exactly which configuration the stars are in), so long as h makes no specific predictions regarding it.

This is related to the converse tactic of adding *ad hoc* elements to a theory (discussed in chapter 3, page 80), for example, the probability that e ('Jones was shot five times in the head') given h ("Jones was murdered") remains virtually 100 percent regardless of who committed the murder. Any more specific theory such as "Smith murdered Jones" would not reduce that probability. It would remain virtually 100 percent (assuming there is no other pertinent evidence). But specifying such a theory *will* reduce the prior probability, because that must be divided among all possible suspects (see my example of assassination theories on page 227). But this is only because specifying the murderer is now a component of *h*. In the case of contingencies like whether Smith or Jones observed the *e* that was predicted by *h*, the specification of who would make the observation is *not* a component of *h*, and therefore no reduction of the prior probability ensues (for exactly the reasons already explained in chapter 3: the prior probability that it was "Smith or Jones" equals the sum of the prior probabilities of every relevant possibility, e.g., "only Smith," "only Jones," and "Smith and Jones"; yet *h* predicts only "Smith or Jones," which includes all of the above). Yet the consequent probability *also* remains the same. So this contingency has no effect at all.

Thus, per the example I used in chapter 3 (page 77), if *h* is a theory of the origins of Christianity that makes no predictions regarding which exact name the Gospel of Mark would be assigned, but only predicts that it would be assigned some name (or indeed, doesn't even entail anything about whether it would or wouldn't be named at all, only that it would be written), then we don't have to concern ourselves with the probability that the name would be Mark. And this follows all the way down the line, for example, *h* doesn't even have to predict specifically that that Gospel would be written, but only that some sacred story about Jesus would have come to exist conveying at least the information *h* entails was paramount to early Christians, which prediction could be satisfied by a largely different text than was actually produced. Also, sometimes even when *h* does make one name more likely than another (or one text more likely than another, or anything more likely than another), it may do so to such a slight degree (warranting only a small difference in probability either way) that the difference is washed out by *a fortiori* estimates and thus can be ignored anyway (this phenomenon of *a fortiori* estimates "washing out" small probabilities was explained in chapter 3, page 85).

This has broad epistemological importance. Just as 'exactly by whom and exactly when' is not normally a predictive component of *h* in any scientific experiment, so 'exactly what' is not normally a predictive component of *h* in any historical theory. This goes beyond irrelevancies entailed by our background knowledge. For example, we can formulate an *h* about Jesus that only entails three specific things would be said about him, regardless of how, in what medium, or in conjunction with what else. Such an h renders the appearance of those three things highly probable in any surviving text about Jesus from that period, of whatever sort, without entailing anything else about what those texts would be or say. Thus, we needn't calculate the odds that the Gospel of Mark would be produced, word for word, exactly as we have it (which odds would be astronomically small, given all the possible configurations of words that could convey the same things). No hypothesis usually makes any predictions regarding such specifics, and thus any coefficient of contingency accounting for them would identically affect both consequents, and thus would mathematically cancel out. Only, of course, if a hypothesis *did* make a differential prediction regarding such details would the coefficient of contingency entailed have to be accounted for, and even then only if the difference was large enough to matter. The method of emulation criteria (analyzed at the end of chapter 5, page 192) is in effect a miniature example of that, where one hypothesis proposes certain components of a text are there by chance (either the chance decisions of the author or the chance coincidences of history), but another proposes those components are there by design (comparable to the stars being in that eerie cross pattern suggested earlier). But even then (as I discussed before) a hypothesis of design rarely makes exact predictions, but rather generic predictions that are merely more exact than a hypothesis of chance would entail. In other words, like the Big Bang Theory and the arrangement of the stars, you can formulate a hypothesis that only makes predictions regarding general characteristics. This is just as true in science. For example, a scientific hypothesis can predict the general pattern of events to expect from a volcanic eruption, without predicting exactly what will happen (such as which direction ash will drift, because that will be contingent on factors unrelated to volcanoes, such as the prevailing wind at the time).

But sometimes the solution pertains to the role of b, rather than the structure of h. In chapter 3 I began by comparing two examples—a hypothetical darkness in 1983 and a purported darkness in the 30s CE—and one factor that came up was the fact that so much evidence survives from 1983, whereas evidence that survives from antiquity has passed through a highly destructive and largely random filter. The hypothesis h "Jesus was executed by Pontius Pilate" (in conjunction with our background knowledge b) entails that an official record of the trial and verdict was created and filed in the Roman archives of Caesarea. If Jesus had been tried and executed in New York in 1983, the same would have occurred. Yet if we pored through court archives from 1983 and found no trace of that trial, this would substantially lower P(e|h.b), because h predicts evidence that didn't turn up. Yet surely not having the same court record from the time of Pilate shouldn't lower P(e|h.b) at all, because (given our background knowledge) we have no reason to expect *that* record to survive. And yet h does not entail the prediction "an official Roman record of the trial would not survive," that is, h does not *predict* the absence of that court record. Indeed, if next year we found a stash of official first-century documents buried under modern Caesarea that included exactly that trial record (and the find was fully authenticated), we would count this as greatly *increasing* the probability that h is true (so enormously, I suspect, that h would become an unassailable certainty). And yet h cannot be stated as predicting we would *find* that record, because then our *not* having found that record would have to reduce P(e|h.b), indeed quite substantially (in fact, by exactly as much as *having* that record would increase it).

This problem ranges far beyond this one example. Nearly every hypothesis about antiquity entails the existence of vast quantities of evidence next to none of which survives or was even expected to have survived—or in fact none survives at all, and wasn't expected to. But this already follows from the fact that consequent probabilities are conditional on both h and b, and b entails our knowledge of the scarce and random survival of ancient evidence like this. So the difficulty is easily resolved in the logic of BT. The probability that Pilate's "record of the trial" would survive is small, due to the contents of b, but that does not mean P(RECORD IS FOUND|h) is small (and therefore finding it would reduce (!) the epistemic probability of h). That's because h entails P(RECORD IS FOUND|h) is very high only *given the record's survival* (through the usual destructive filter all ancient evidence has passed through), and the outcome of that contingency is not entailed by h (as long as h makes no prediction whether that record will have survived that filter), so if the record turns up (and thus survived the filter after all), its discovery should still increase the epistemic probability of *h* as expected.

Thus, making the consequent probability *also* conditional on *b* is what makes the difference here (hence in BT this probability is in fact P(RECORD IS FOUND|h.b) and not just P(RECORD IS FOUND|h)). So, either the record existed or it didn't; *h* predicts that it did; but *b* entails that even if it did, it probably didn't survive (note that *h* does not entail this, only *b* does—at least in this example). This circumstance is analogous to the 'trustworthy neighbor' example in chapter 3 (page 74). If R = 'the record existed' and F = 'such a record would have been found by now,' then I'll assign these arbitrary numbers just for the sake of argument:

$$P(F|R.b) = 0.01 \; [\ldots which\ entails \ldots] \; P(\sim F|R.b) = 0.99$$
$$P(F|\sim R.b) = 0.001 \; [\ldots which\ entails \ldots] \; P(\sim F|\sim R.b) = 0.999$$

You might think P(F|~R.b) = 0, since if it didn't exist, obviously it won't have been found, but there is a nonzero probability of forgeries and erroneously filed records. In other words, just because we find such a record does not automatically entail *h* is true, because the record we find may be a forgery or may have been filed erroneously in the first place (and of course there are all the more extreme possibilities, such as that we're hallucinating our finding the record—but those usually have vanishingly small probabilities). Thus I assign P(F|~R.b) = 0.001 to reflect these possibilities (though to have such a low probability requires the record to survive a reliable process of authentication, since forgery is so common, particularly in the field of biblical antiquities).

It's also true that apart from the filter, I'm assuming P(R|h.b) equals one, even though usually it will be something less than one, for example, there is always some small probability that an official record that would usually be made and filed didn't get made or filed, but that probability is often small enough to ignore. Likewise, if there is any chance we would know, if the record didn't exist, that it didn't exist (as sometimes is the case, e.g., there is always some small probability that someone in antiquity would have checked and reported it didn't exist, and that report could have survived), that would also have to be factored in, but again this probability may be so small that it can be ignored (see my analysis of the Argument from Silence in chapter 4, page 117). Conversely, it's also possible to have such a report about the record that is *itself* a lie or in error, thus even having a report of the record's nonexistence would still not strictly entail the record didn't exist, requiring an estimate of probabilities again (likewise for a report that claimed it did exist). I will ignore all these possibilities here (and assume instead that h strictly entails R and that $\sim h$ makes no predictions other than $\sim$R). But I make a point of noting all this here because sometimes such factors will have a large enough effect that they cannot be ignored.

You might then object to my assignment of $P(\sim F|\sim R.b) = 0.999$ since if the record didn't exist, our not having it is still not so certain—for we could have turned up a forgery or an administrative error by now. But $P(\sim F|\sim R.b)$ must reflect the actual probability of a forgery or administrative error— which is *not* their probability given that the document is found, since given that the document *isn't* found the probability that we should still *expect* such a document to have been erroneously or deceitfully produced by now is not the same. In fact, the latter is usually vanishingly small. Hence I could even assign $P(\sim F|\sim R.b) = 1$ as a practical stand-in for $P(\sim F|\sim R.b)$ $\rightarrow 1$, which reflects the assumption that such forgeries and errors are rare enough that we shouldn't ever *expect* them to exist (as if every bogus item of evidence conceivable had been forged by now)—whereas we do have grounds to suspect forgery when a suspiciously convenient document actually turns up, and for that reason I did not allow $P(F|\sim R.b) = 0$. Technically this forbids $P(\sim F|\sim R.b) = 1$, since it is necessarily the case that $P(\sim F|\sim R.b)$ $= [1 - P(F|\sim R.b)]$, and therefore must be 0.999 if $P(F|\sim R.b) = 0.001$ (since given $\sim$R the alternatives F and $\sim$F exhaust all possibilities, and therefore their respective probabilities must sum to 1). But the difference between 1 and 0.999 is too small to matter in the present case (whereas the difference between 0 and 0.001, in fact any nonzero number, is effectively infinite). So I will use 1 only to simplify the math, because that won't change the outcome in any visible way.

Given these numbers, then a Bayesian analysis that hinged solely on this piece of evidence would go as follows. If the record is not found (the state of evidence we are actually in) and if h entails R and (let's say) P(h|b) = 0.9 (i.e., we are otherwise convinced h is probably true) then:

$$P(h|e.b) = \frac{P(R|b) \times P(\sim F|R.b)}{[\,P(R|b) \times P(\sim F|R.b)\,] + [\,P(\sim R|b) \times P(\sim F|\sim R.b)\,]}$$

$$P(h|e.b) = \frac{0.9 \times 0.99}{[\,0.9 \times 0.99\,] + [\,0.1 \times 1\,]}$$

$$P(h|e.b) = \frac{0.891}{0.891 + 0.1} = \frac{0.891}{0.991} = 0.899$$

So the absence of the record *does* reduce P(h|e.b), from an initial belief of 0.900 to a revised belief of 0.899, but this change is so little as to make no practical difference. In fact, since P(F|R.b) should really in this case be far lower than 0.01 (i.e., the probability we'd have Pilate's court records is surely far less than one in a hundred, indeed probably less than one in a million), the effect of the missing evidence is really even smaller, in fact so small as to be effectively invisible. Hence we can ignore it. So our intuition that the absence of this evidence should not lower P(e|h.b) "at all" was technically wrong but in practice correct; we just don't have any convenient vocabulary to express "as near to not lowering it at all as is practically the same as not lowering it at all" so we revert to "not at all" because our intuitions tell us that's close enough. Only when we *don't* have an extremely high expectation the evidence would be lost would that not follow (hence my analysis of the Argument from Silence in chapter 4, page 117).

Meanwhile, if the record *is* found:

$$P(h|e.b) = \frac{P(R|b) \times P(F|R.b)}{[\,P(R|b) \times P(F|R.b)\,] \; + \; [\,P(\sim R|b) \times P(F|\sim R.b)\,]}$$

$$P(h|e.b) \; = \; \frac{0.9 \times 0.01}{[\,0.9 \times 0.01\,] + [\,0.1 \times 0.001\,]}$$

$$P(h|e.b) \; = \; \frac{0.009}{0.009 + 0.0001} \; = \; \frac{0.009}{0.0091} \; = \; 0.989$$

Thus finding the record *does* increase P(h|e.b), exactly as expected, from an initial belief of 0.9 to a revised belief of 0.989, representing a rather large increase in our confidence that *h* is true. Which is all as we intuited should be the case.

These analyses can be repeated for any other comparable case, where we can't predict from *h* exactly what evidence we would now have, either due to *h* itself entailing nothing either way (except at most in some general respects) or due to *b* entailing a change of expectations from what they'd be given *h* alone. Thus, that being a fact does not impair historical reasoning at all, and any philosophers or mathematicians who've ever worried about this can rest easy. For example, McCullagh discusses an example in which a very different type of contingency played a role: an event occurred in a private household, which just happened to be witnessed and recorded in a diary by a traveling Frenchman, allowing us now to argue that the event occurred by appealing to the evidence of his diary.[12] And yet our *h* ("the event occurred") in no way predicts that there would have been a traveling Frenchman just happening by. That is actually very improbable, hence our evidence is very improbable. Yet surely its being improbable should not lower the probability of *h*. To the contrary, such evidence should increase that probability, quite substantially in fact. In other words, if M = 'that Frenchman happened by and wrote in his diary what he saw' and V = 'that event happened,' then (all else being equal) P(V|M) should be high. But h only predicts V, not M, while b entails M is very improbable. Here we'll assume h entails V, so in BT we are only concerned with determining P(V|e.b), and if we assume M constitutes e (we actually have the Frenchman's diary, and that's all we have), then P(V|e.b) = P(V|M.b), which produces:

$$P(V|M.b) = \frac{P(V|b) \times P(M|V.b)}{[\,P(V|b) \times P(M|V.b)\,] + [\,P(\sim V|b) \times P(M|\sim V.b)\,]}$$

If we split M into D and F, D = 'diary entry attesting V' and F = 'Frenchman happened by,' and assume P(V|b) = 0.5 (i.e., we have no prior reason to suspect V either did or didn't happen), and that the Frenchman happening by has a contingency coefficient of n (representing the improbability of that coincidence, i.e., P(F|b) = n), and if we assume that P(D|V.F.b) = 0.99 (i.e., if when we assume F and V, then the odds we'd have the diary entry are nearly 1; yes, the further contingency of the Frenchman, given his being there, recording the event, could similarly be analyzed, but that would take the same form as the contingency of the Frenchman being there, and thus that analysis would look essentially the same as the following) and if we assume that P(D|~V.F) = 0.01 (i.e., if when we assume F and ~V, then the odds we'd have the diary entry are 0.01, i.e., the small probability the Frenchman would make the story up), and if we assume, of course, that if ~F, then ~D (so both P(D|~V.~F.b) and P(D|V.~F.b) = 0, i.e., if the Frenchman didn't happen by, the diary entry wouldn't exist, and therefore ~M), then:

$$P(M|V.b) = [\,P(D|V.F.b) \times P(F|b)\,] + [\,P(D|V.\sim F.b) \times P(\sim F|b)\,]$$

$$= [\,0.99 \times n\,] + [\,0 \times (1-n)\,]$$

$$= 0.99n + 0$$

$$= 0.99n$$

$$P(M|\sim V.b) = [\, P(D|\sim V.F.b) \times P(F|b) \,] + [\, P(D|\sim V.\sim F.b) \times P(\sim F|b) \,]$$

$$= [\, 0.01 \times n \,] + [\, 0 \times (1 - n) \,]$$

$$= 0.01n + 0$$

$$= 0.01n$$

So:

$$P(V|M.b) = \frac{0.5 \times 0.99n}{[\, 0.5 \times 0.99n \,] + [\, 0.5 \times 0.01n \,]}$$

$$P(V|M.b) = \frac{0.495n}{0.495n + 0.005n}$$

$$P(V|M.b) = \frac{(0.495)n}{(0.495 + 0.005)n}$$

$$P(V|M.b) = \frac{0.495}{0.5} = 0.99$$

As before, the coefficient of contingency cancels out and thus disappears, making no difference to the outcome. As expected, given the assigned probabilities, the existence of the diary entry greatly increases the probability that the event happened. The extreme improbability of a Frenchman just happening by is completely moot.

Other concerns about contingency are already resolved by probability theory. For example, sometimes it's claimed that the probability of life arising on earth is very small, whereas if it was by design, the odds would

be very high, creating such an enormous disparity in consequent probabilities that unless you have a wildly outrageous bias against the existence
of a Creator (resulting in an extraordinarily large disparity in the priors
against it), BT entails life was created by intelligent design. But there are
two fallacies in this argument. The first is of invalidly predicting e from h,
when in fact from the hypothesis "God exists" it isn't possible to deduce
the prediction 'simple, single-celled carbon-coded life forms would arise
on just this one planet out of trillions, and only billions of years after the
universe formed, which would only slowly evolve into humans after bil-
lions of years more' etc. Thus P(e|GOD.b) in this instance is not 'very high.'
In fact, arguably it's extraordinarily low, even before adding any background knowledge that renders such divine beings improbable in and of
themselves.[13] The second fallacy, however, is a common mistake in reasoning about probability: the odds of life forming by chance are *not* the
odds of life forming by chance specifically here on earth, but the odds
of life forming by chance on *some* planet *somewhere* in the whole of the
known universe.[14] Because, obviously, wherever that happens to be will
become "specifically here" for whoever ends up evolving on that planet to
think about it. It's the difference between you winning the lottery (which
is very improbable) and someone winning the lottery (which is very probable). You are reasoning fallaciously if, after winning, you conclude the
lottery must be rigged simply because your winning was so very improbable. Because someone was likely to win, and that someone was as likely
to be you as anyone else playing. Hence, in fact, the number of planets and
years available are such that, where L = 'life as we observe it to be' and U
= 'the universe as we observe it to be,' P(L|U)→1. And since (as suggested
earlier), P(L|GOD)→0, the consequent probabilities are in fact exactly the
reverse of what was thought, such that even if P(GOD|b) were high (and it's
not), life *still* probably wasn't created by intelligent design.[15]

The relevance of this to history is that the same kind of fallacious arguments can arise if you do not attend to the correct probabilities. For example,
you cannot argue that Alexander the Great assassinated his father Phillip
because the odds of that assassination happening by chance are small, but
the odds of that happening "if Alexander did it" approach certainty. To

begin with, such coincidences happen all the time (often kings are assassinated who just by chance have sons or successors who will benefit; indeed, this is probably true in most cases)—so the probability that this is one of those coincidences is actually high, not low (I'll discuss this phenomenon using a poker analogy on page 254). But more importantly, in Bayesian analysis this doesn't even become an issue because ~h would have to be accounted for, in which we would list a number of known persons who had the same motive (and that's assuming we can leave out of account the many *unknown* persons who would also have motive), and the prior probability for each being the culprit would have to be the same (assuming we have no other evidence implicating Alexander, or any one else), and, more importantly, the *consequent probability would be the same for all of them.* That is, "Phillip gets assassinated" is 100 percent certain on any "x did it" hypothesis. So Alexander is no more likely to be the culprit than anyone else. In other words, it's fallacious from the start to assume the hypothesis "Alexander did it" is competing against "chance" (as if random quantum events caused kings to be assassinated). Rather, it's competing against other assassins, for every one of whom "the odds of Phillip getting assassinated" are 100 percent. Of course, if we have other evidence, then e is not just "Phillip got assassinated" but the conjunction of all that evidence, which *could* implicate someone specific. Or if there were no other known suspects (or the only known suspects are actually only known from Alexander claiming they are suspects), the prior probability could favor Alexander. If a study of royal assassinations found that, statistically, sons more likely turned out to be the culprit, or that, when there was only one known suspect, more often than not they turned out to be the culprit, such data could be used to alter the priors (if all the contexts are sufficiently similar—see my following discussion of using reference classes to assign priors).

That's just one example. Many more can be imagined. All the scenarios above support the same conclusion: most contingencies can be ignored, and hypotheses can validly make generic predictions exactly as argued in chapter 3. And contingencies that can't be ignored can be fully accounted for in BT.

DETERMINING A REFERENCE CLASS

Strictly speaking, prior probability is the probability of getting a specific kind of *h* when you draw at random from a reference class of all possible *h* → *e* correlations. Those correlations don't have to be causal, although in history they usually are. Because, in history, we are almost always asking what caused *e* and proposing *h* as the answer (see chapters 2 and 3). I'll thus focus mainly on causal hypotheses and explain how to ascertain prior probabilities in a way that can produce intersubjective agreement among expert historians, and when and why such a process is logically valid.

Some critics of BT are skeptical of causal language in applying the theorem, but that's fundamental to many theories, especially historical ones, since any statement about what happened in history reduces to a statement about what caused the evidence we have. And you can't propose historical *explanations* without proposing causes. Historians do distinguish claims about what happened (or once existed) from claims about why it happened (or why it existed). But ultimately all claims about 'what' entail claims about 'why.' For example, we can talk about what the frequency of a particular name actually was in Roman times by talking about the frequency of that name in inscriptions, but that entails assuming a causal relation between actual name frequencies and the appearance of names on inscriptions, whereas merely talking about the frequency of names on extant inscriptions, without any interest in what caused this frequency, is all but useless to a historian, not only because you must assume actual name frequencies is what caused the inscribed name frequencies, but *especially* because even the claim that these inscriptions are ancient entails an unavoidably causal theory about how they came to exist, and for a historian to disregard even the question of whether Roman inscriptions are ancient (or even Roman) is simply an abandonment of history as a field of inquiry.

This remains the case even when the causal relationship appears the other way around. For example, a hypothesis of murder will explain evidence of preparations for that murder, even though the murder didn't "cause" that evidence (since the preparations preceded the murder). Yet the hypothesis still entails there is a causal relationship between the murder and the preparations: in this case, the *intent* to murder, which is inherent in that hypothesis, will have caused both. Similarly, a hypothesis that a religious riot was caused by prior beliefs of that community (such as an ancient prophecy) in conjunction with new events (such as the appearance of a comet) obviously proposes a causal relationship between those prior beliefs and the riot, but not that the riot caused those beliefs. That the prior beliefs existed is evidence supporting the hypothesis (which is that "they rioted because of that ancient prophecy") and therefore this hypothesis makes that evidence more probable, even though the riot did not cause that evidence, but the other way around (the prophecy, in part, caused the riot—the very causal relationship being hypothesized).

Formally speaking, if the riot occurred because of that prophecy, then the probability that there would be no such prophecy (or $P(\sim e|h.b)$) is zero, so the consequent probability (or $P(e|h.b)$) of that item of evidence is 1 (because $P(e|h.b)$ always equals $1 - P(\sim e|h.b)$, a useful observation I'll discuss on page 255). In other words, on that hypothesis, the existence of the prophecy is exactly what we should expect (in fact, if the hypothesis were true, the absence of that evidence would be impossible—apart, of course, from the contingency of that evidence being lost, as we discussed earlier in this chapter). On the other hand, if the riot did not occur because of that prophecy (in other words, if $\sim h$), then the probability that there would be no such prophecy (or $P(\sim e|\sim h.b)$) is not zero, and therefore the consequent probability (or $P(e|\sim h.b)$) of that item of evidence is *less* than 1. Because then, it is not *exactly* what we would expect to be in evidence (even if it's not wholly unexpected). Of course, if $\sim h$ is the hypothesis that the *absence* of the prophecy caused the riot, then $P(e|\sim h.b)$ is not only less than 1 but in fact nearly zero, since the prophecy is in evidence, and that is exactly the opposite of what that hypothesis predicts (the consequent only escapes *not* being zero because of such possibilities as that the prophecy existed but no one knew about it or that the prophecy was fabricated after the riot). But still we are talking about a causal hypothesis, whether it's a hypothesized event causing the evidence, or the evidence causing the hypothesized event. Or, as in the case of name frequencies on inscriptions, we are talking about a fact *assumed* by a hypothesis causing the evidence:

that a particular frequency of names in ancient Rome caused the frequency of names on surviving inscriptions.

Thus all historical claims that Bayes's Theorem can ever test must involve causal hypotheses, which link the claim to the evidence. But those causal hypotheses need not always be fully specified. For example, if (let's say, with a sample size in the hundreds) archaeology confirmed 8 out of 10 Roman colonies (cities established by the Roman government for settling war veterans) had public libraries, we could use that as a prior probability that a newly excavated Roman colony had a public library, without specifying the exact causal relation that produced this probability. We are still implicitly assuming there is one, that is, *something* caused Romans to regularly fund public libraries in their colonies and *something* caused them not to from time to time, but we don't need to know what either "something" was, since whatever those "somethings" were, we already know what frequencies of outcome they generated, and that's all we need in this case. Certainly, if we acquire information regarding those causes, then that information becomes relevant again (hence it's worth repeating here that the results of BT are always conditional on current knowledge—when we get new information, those results may change, the epistemological significance of which I'll discuss later on page 276). For example, if we discovered that certain specific families were responsible for funding most of those libraries, we might be able to revise our probabilities accordingly. If what caused most of those libraries was the patronage of those wealthy families, such that all the identified colonies that received their patronage had libraries, and only 2 out of 10 other colonies had libraries (let's say, because the veterans settled in those 2 in 10 cases had the means and interest to combine their own resources to establish a library to bring more prestige to their colony), then if the colony we are newly excavating can be independently determined to have received patronage from those identified families or not, our priors can be revised. If our city is determined to have had that special patronage, and our data shows 67 out of 67 cities with their patronage have public libraries, then the prior probability our new city did as well will now be over 98%.[16] Of course that assumes the evidence establishing this city had their patronage is so strong that the probability of that connection is extremely high, high enough that the odds of our being wrong about it make no discernible difference to the result. But that aside, what we get is 98%, which is a lot higher than the 80% we were working with before. Hence with better information we get better estimates. Likewise, if our city is determined *not* to have had that special patronage, and assuming all else is the same, then the prior probability our new city had a public library will only be about 20% (2 in 10), or perhaps closer to 21% (see note 16, page 326). Again, better information, better estimate.

This is what is called a **reference class**. In that last case, the reference class is 'Roman colonies lacking special patronage,' whereas in the preceding case the reference class is 'Roman colonies *receiving* special patronage,' which two classes are mutually exclusive, but both comprise a larger reference class of just 'Roman colonies.' If all we know is that a city we are excavating falls in that last (combined) reference class, then we must use the prior probability that that class entails. Only if we know it falls into one of those more specific sub-classes can we use the prior probabilities that *those* classes entail. The challenge of ascertaining prior probability always reduces to this same exercise of determining the most relevant reference class (for which we have, or can hypothetically construct, credible frequency data). Imagine you can put all cases into a hat (even the ones we don't know about yet) and scramble them up and then draw one of them from that hat at random. The prior probability that a new case will exhibit the relevant feature corresponding to h equals the chance of drawing it out of that hat. For example, following the previous example, if we put all Roman colonies into a hat and drew one at random, we'll draw a city with a public library out of that hat an average of 8 out of every 10 pulls, making the chance of such a draw 80%. Hence the prior probability is 80%.

Many object that there is rarely any objective way to settle on how to determine the prior probability, because any given hypothesis will simultaneously belong to countless reference classes, and which reference class you use to develop the prior can seem rather arbitrary.[17] But when two parties come up with different ways to determine the prior probability (because, as John Earman says, "there are different ways of conceptualizing an inference problem" in Bayes's Theorem, just as in any other method), we must ask whether party A's prior is based on more or less information than party B's. Is A's reference class narrower or better understood than B's? If the answer is yes, then party A's prior is to be preferred, and vice versa. To do otherwise would be to willfully ignore information, as if we know a Roman colony had special patronage, yet used the broader reference class of 'all Roman colonies' anyway. That's a violation of the logic of BT, which requires the prior probability to be conditional on b, and the fact that our case falls into a narrower reference class is information in b. In effect, we *know* the prior probability in this case is 20% and not 80%, and BT entails we must use what we know. I call this the **rule of greater knowledge**. When we know more, our estimates must reflect that greater knowledge. We can't pretend we don't have it.

If, on the other hand, the two competing reference classes are epistemically equal, and we don't know what's in the sub-class that is a conjunction of those two competing classes, then the problem of selecting the reference class gets more complex. For example, suppose we are excavating a newly discovered Roman colony in Italy named Seguntium, and we want to argue that a building we've uncovered is a public library, and we have our consequent probabilities worked out, but we need to determine our priors. Suppose, also, that we already know to a very high probability that it is such a colony and it did not have special patronage, and that we also know colonies in Italy had public libraries at a rate of 90% instead of the 80% rate more generally. We now have two different reference classes, each giving wildly different estimates of prior probability: the 'no patronage' class at 20% and the 'in Italy' class of 90%. Ordinarily, of course, with the kind of data imagined for this scenario, the frequency of public libraries in the reference class 'Roman colonies in Italy without special patronage' would also be known, and would necessarily supersede all others (per the rule of greater knowledge, establishing the logical requirement of preferring the narrowest available sub-class, and here we would have a proper sub-class produced by the conjunction of the competing classes 'without patronage' and 'in Italy'). But it's possible we never found adequate information to establish or rule out the 'special patronage' in any of the other Italian colonies, and thus we can't determine the statistical content of the reference class 'Roman colonies in Italy without special patronage.' We're then stuck with two equally applicable reference classes that give entirely contradictory indications of the prior probability.

In such a circumstance, the simplest solution might be to generate a conclusion with an upper and lower bound, thereby using both reference classes (the 'no patronage' class giving us the lower bound and the 'in Italy' class giving us the upper bound). If there were many more simultaneously competing classes, one would use the two that entail the highest and lowest priors since the resulting span will thus encompass all the others anyway. But that solution is often incorrect (as the true prior could well be outside the resulting range) or renders results too ambiguous to be of any use. Usually we should prefer one class over the other (and in this example we should, as I'll explain on page 238).

In reality, of course, there still *is* a sub-class 'Roman colonies in Italy without special patronage'; it's just that in this example we don't know what's in that class, forcing us to guess, thus introducing a wider margin of error. For example, the difference between the Italy class and the more general class ('all Roman colonies') of 90% rather than 80% having libraries, may be entirely the result of more colonies in Italy receiving that special patronage than elsewhere (in which case the reference class 'colonies without patronage' would be the more accurate class and our prior should be 20% and not 90%). But it might also be the result of veteran settlers in Italy being wealthier than elsewhere and thus funding more libraries on their own, in which case the reference class 'all colonies without patronage' would be the wrong class, because the narrower 'colonies without patronage in Italy' entails a different prior probability, perhaps 30% instead of 20%, in any case some higher frequency presently unknown to us due to our lack of information. Or there could be any of countless other possibilities. Of course, just as in the case of 'all colonies' vs. 'colonies without patronage,' we should prefer the more general class when we don't know the frequency in the sub-class, so we should here, too, and thus prefer the more general class 'colonies without patronage' until we know more about the sub-class 'colonies without patronage in Italy.' I'll say more about why later (page 238). But until we have actual information regarding such possibilities, we may have to accept the huge range of uncertainty entailed by the two reference classes we *can* identify but can't pare down.

And for all that, it's still possible the conjoined class entails a *lower* or *higher* prior probability; for example, if for some reason all 'colonies without patronage in Italy' have no libraries, or if for some reason all of them do (or any other frequency). It's just that we have no information that makes either likely. That is, given the information we have, we should expect 2 in 10 'colonies without patronage' to have libraries, not (for instance) 10 in 10, or 0 in 10, even in Italy. In other words, until we know otherwise we have to assume the frequency for the whole region applies to each sub-region. The mere possibility that things could be different in Italy does not warrant assuming they are (hence my fifth axiom in chapter 2, page 26). We might later be able to prove otherwise, but until then we must base our assumptions on what we *now* know. This is logically required by BT. And when our knowledge indicates two different possibilities, we have to allow either to be the case until we can narrow it down (and as it happens, in this case we can, as I'll show on page 238). And, of course, if we also rely on *a fortiori* estimates we'll be on even safer ground (see chapter 3, page 85).

The following Venn diagrams illustrate the four most common conditions of competing reference classes:

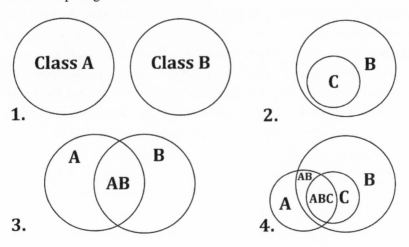

Condition 1 can never produce a valid conflict. If two reference classes actually apply to the same hypothesis, that fact logically entails their conjunction (since at the very least, their conjunction will contain our hypothesis—provided our hypothesis is logically possible). Thus, when faced with Condition 1, we need only ascertain which reference class applies to our h, A or B. Condition 2 is more typical, a case of narrowing the reference class. If our hypothesis resides in Class C and we can derive a prior probability from that class, we must do so. Because knowing h resides in C constitutes more information about h than is entailed by B alone. Condition 3 is also common, and similar to Condition 2. If our hypothesis resides in Class AB and we can derive a prior probability from that class, we must do so. Because knowing h resides in AB constitutes more information about h than is entailed by either A or B alone. If, however, we cannot derive a prior from that sub-class (because we lack the requisite data for it), and the priors entailed by A and B differ, we *can* conclude the prior *probably* falls somewhere in between (unless we have definite knowledge already that their conjunction would change that expectation, e.g., carbon, sulphur, and potassium nitrate each have low probabilities of catching fire, but their conjunction has an extremely high probability of that, thus the combined class takes on properties well outside the average of the individual classes due to the causal interaction of the parts—and as in chemistry, sometimes also in history and social systems). But we can often get more specific than that. As I'll explain in a moment, there are some logical shortcuts we can take to show that the prior more probably falls nearer one side than the other (we can even apply sophisticated techniques in probability calculus on the complete set of data to get essentially the same results, but that's beyond the scope of the present book). Finally, Condition 4, exemplifying a more complicated case, simply combines the circumstances of Conditions 2 and 3.

The actual statistical problem created by the Seguntium example could become very complex and might have to take into account many other variables. Historians rarely face such problems, and even more rarely have the skill set to solve them (although they can always collaborate with mathematicians, and arguably sometimes should). But historians do need some simple rules of their own for rationally negotiating complex cases that have little or no exact data. To illustrate this, the libraries scenario can be represented with this Venn diagram:

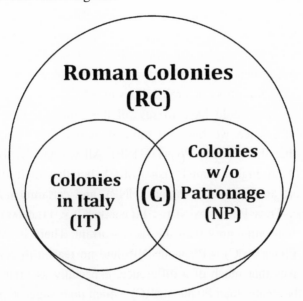

In this example, P(LIBRARY|RC) = 0.80, P(LIBRARY|IT) = 0.90, and P(LIBRARY|NP) = 0.20. What's unknown is P(LIBRARY|C), the frequency of libraries at the conjunction of all three sets. If we use the shortcut of assigning P(LIBRARY|C) the value of P(LIBRARY|NP) < P(LIBRARY|C) < P(LIBRARY|IT), that is, P(LIBRARY|C) can be any value from P(LIBRARY|NP) to P(LIBRARY|IT), then the first concern is how likely it is that P(LIBRARY|C) might actually be less than P(LIBRARY|NP), or more than P(LIBRARY|IT), and the second concern is whether we can instead narrow the range. Given that we know Seguntium lacked special patronage, in order for P(LIBRARY|C) < P(LIBRARY|NP), there have to be regionally pervasive differences in the means and motives of veteran settlers in Italy—enough to make a significant difference from veteran settlers in the rest of the Roman empire. And indeed, on the other side of the equation, for P(LIBRARY|C) > P(LIBRARY|IT) these deviations would have to be remarkably extreme, not only because P(LIBRARY|IT) > P(LIBRARY|RC), but also because P(LIBRARY|RC) is already >>

P(LIBRARY|NP), which to overcome requires something extremely unusual. Lacking evidence of such differences, we must assume there are none until we know otherwise, and even becoming aware of such differences, we must only allow those differences to have realistic effects (e.g., evidence of a small difference in conditions cannot normally warrant a huge difference in outcome; and if you propose something abnormal, you have to argue for it from pertinent evidence—which all constitutes attending to the contents of *b* and its conditional effect on probabilities in BT).

However, we would have to say all the same for P(LIBRARY|C) > P(LIBRARY|NP), since we have no more evidence that P(LIBRARY|C) is anything other than exactly P(LIBRARY|NP). All we have is the fact that P(LIBRARY|IT) is higher than P(LIBRARY|RC), but that in itself does not even suggest an increase in P(LIBRARY|NP), and certainly not much of an increase. Thus P(LIBRARY|NP) < P(LIBRARY|C) < P(LIBRARY|IT) introduces far more ambiguity than the facts warrant. There is every reason to believe P(LIBRARY|C) ≈ P(LIBRARY|NP) and no reason to believe being in Italy makes that much of a difference, especially as P(LIBRARY|IT) is only slightly greater than P(LIBRARY|RC), which *does* suggest only a small rather than a large difference between Italy and the rest of the empire, and likewise we should expect the large disparity between P(LIBRARY|NP) and P(LIBRARY|RC) to be preserved between P(LIBRARY|C) and P(LIBRARY|IT), as the causes producing the first disparity should be similarly operating to produce the second—unless, again, we have evidence otherwise. In short, NP appears to be far more relevant a reference class than IT in this case and should be preferred until we know otherwise. And if we also use *a fortiori* values (setting the probability at, say, 10–30%), we will almost certainly be right to a high degree of probability. All this constitutes a more complex application of the rule of greater knowledge. When you have competing reference classes entailing a higher and a lower prior, if you have no information indicating one prior is closer to the actual (but unknown) prior, then you must accept a margin of error encompassing both, but when you have information indicating the actual prior is most probably nearer to one than the other, you must conclude that it is (because, so far as you know, it is). In short, we can already conclude that it's so unlikely that P(LIBRARY|C) deviates by any significant amount from P(LIBRARY|NP) that we must conclude, more probably than not, P(LIBRARY|C) ≈ P(LIBRARY|NP), regardless of the difference between P(LIBRARY|IT) and P(LIBRARY|RC). And as in this case, so in many others you'll encounter.[18]

All the same follows even when such precise and abundant data is not available to determine priors. We often have to assess a theory's relative plausibility subjectively in light of a kind of holistic polling of our background knowledge (the logic of which I'll discuss in the next section, starting on page 257). If we stick to *a fortiori* reasoning, and do our best to ensure we are honest and discerning when polling our background knowledge, and as long as we are as well informed as we reasonably can be in the circumstances, then this will still produce better-than-arbitrary results, in fact often entirely reasonable and defensible results. After all, that's why any knowledge of the past is possible. It's also why expert opinion carries greater weight (as argued in chapter 2): as far as analyzing claims in their own field, experts have seen more relevant data and thus can get more informed results when holistically polling their past experience. But this still requires actual confirmed experience, not past hunches. In other words, it requires data that can be communicated, shared, or repeated by others, which means ultimately an expert must be able to adduce many actual examples confirming his statistical opinions in general when called upon to justify his estimates, and if he cannot, then his estimates are not justified or are too weakly justified to carry any special weight.

Sometimes a competing reference class becomes moot. As noted in chapter 5 (page 168), we can run a series of BT arguments by starting with a neutral prior (0.5) and then run a single case, then run another case using the outcome of the first case (i.e., its posterior probability) as the prior in the second case, and so on down the line until we've exhausted all known cases capable of analysis. The end result will be the correct Bayesian conclusion (the correct epistemic probability of *h* in light of all evidence *e* and all background knowledge *b*).[19] And sometimes when there are two competing reference classes, both of them will get picked up eventually in this series, and thus it won't matter which one we start with (since mathematically the outcome will be the same, just as it doesn't matter whether we multiply 6 × 5 or 5 × 6, you still always get 30). This is not the case for the Seguntium example (because the competing classes in that case are not equal, i.e., upon analysis only one of them was found likely to be close to the actual reference class). But it can happen. We could divide any body of evidence into two and derive a prior probability from one of them and use the other to develop the consequent probabilities. In such a case, it won't matter which one we use for which; the outcome must logically be the same (and if it isn't, we've erred in our math somewhere).

For example, we could use features in the Gospels that place Jesus in a particular reference class, like 'legendary rabbis,' and derive a prior probability from that (if we had enough data to construct that class). The number of legendary rabbis that happen to be fabricated in ratio to all legendary rabbis (fabricated *and* historical) would then equal (more or less) the prior probability that Jesus was fabricated, too. For in determining the prior, we must treat all members of the class the same (e.g., *prior* to considering our specific evidence *e*, Jesus is as likely to have been fabricated as any other legendary rabbi). If we actually had information that allowed us to treat some members differently, then that would entail we have a viable sub-class and should use *that* instead. For example, if we knew that 'fabricated rabbis spoken of within a generation of their alleged lifetime' entailed a much smaller ratio of fabricated-to-real rabbis than the broader class of all 'fabricated legendary rabbis' does, then that could greatly reduce the prior. Of course, in practice we rarely have the information necessary to construct that sub-class. We almost never know when a 'fabricated rabbi' was fabricated relative to the period in which he is said to have lived (since the earliest evidence of stories about them almost never survives). And since priors must be based on what we know (not on what we merely suspect), we cannot get that precise, and thus must use the broader class. (It shouldn't need repeating that if we based the priors on what we merely suspected rather than what we actually know, this qualification commutes to the conclusion, and thus the output of our whole analysis will only be 'what we suspect' and not 'what we know,' and if the latter is what you want, knowing the former is useless.) It might still be possible to appeal to other data pertaining to the timeline of legendary fabrication of historical personages generally in order to develop a refined prior from the rabbi set specifically, along lines similar to the weighting of the competing priors in the Seguntium case. But at any rate, from an actual or reconstructed reference class of 'legendary rabbis' we can begin our analysis.

Or we could start with a completely different prior, based on oddities in the letters of Paul. If it's true that the contents of those letters are bizarrely silent about a historical Jesus (that's debatable, but let's assume it for the sake of argument), then you will have a hypothetical reference class of 'bizarrely silent letters about supposedly historical persons,' in which the number of cases where that bizarre silence is caused by the person's non-existence can be set in ratio to the number of all cases (those and all others, the other cases being those in which such a bizarre silence has other causes instead). Of course, we have so little data to reconstruct that class it might not merit preference (and the conjunction of that class with the 'legendary rabbis' class might have only one member: Jesus). If we have more and better data for the 'legendary rabbis' class, by the rule of greater knowledge, we must use the latter to determine the prior. But then the oddities of the letters would still enter the contents of e, affecting the consequents. In fact, they must affect them in mathematically the exact same way (see following notes for a demonstration). Hence the rule of greater knowledge would be moot in this case. Nevertheless, applying the rule of greater knowledge is the wiser tack, since the subjectivity of priors is one of the main sticking points in debates over BT conclusions, so it's always best to use the most objectively determinable prior possible. But even doing that here, we still must estimate the *consequent* probabilities, that is, P(SILENCE OF THE EPISTLES|JESUS EXISTED.b) and P(SILENCE OF THE EPISTLES|JESUS DID NOT EXIST.b). Yet from all the same data we could mathematically construct a specific reference class and derive a *prior* probability that would mathematically alter the outcome of BT by exactly the same amount. But that would require complicated calculations that are far too unnecessary. This is especially true if we are using *a fortiori* estimates, as then that required mathematical agreement will be subsumed by our margins of error anyway and we needn't worry about it.

It might not be clear why. But if our consequents were P(SILENCE OF

THE EPISTLES|JESUS EXISTED.b) = 0.2 and P(SILENCE OF THE EPISTLES|JESUS DID NOT EXIST.b) = 0.6, this would convert to a prior probability of P(JESUS EXISTED|b) = (0.5 × 0.2) / (0.5 × 0.2) + (0.5 × 0.6) = 0.25. But that's only if we use a neutral initial prior (of 0.5), which requires us to move the knowledge we have about legendary rabbis into e and derive new consequents by similarly deconstructing that reference class, i.e., if the 'legendary rabbis' reference class entailed P(JESUS EXISTED|b) = 0.8, then, when likewise beginning with a neutral prior, P(JESUS WAS A LEGENDARY RABBI|JESUS EXISTED.b) and P(JESUS WAS A LEGENDARY RABBI|JESUS DID NOT EXIST.b) must be in such a ratio to each other as to entail (0.5 × P(JESUS WAS A LEGENDARY RABBI|JESUS EXISTED.b)) / (0.5 × P(JESUS WAS A LEGENDARY RABBI|JESUS EXISTED.b)) + (0.5 × P(JESUS WAS A LEGENDARY RABBI|JESUS DID NOT EXIST.b)) = 0.8. Applications of BT, to remain formally valid, cannot violate these conditions.[20] But these mathematical relationships are very complex. If we rely on *a fortiori* estimates, we won't violate this consistency requirement and thus needn't worry about it. And the latter is a lot easier to do than running complex mathematical tests for consistency, especially for historians (although, it's worth pointing out, running such a test can sometimes be a viable means of demonstrating that a historian is employing inconsistent probability estimates).[21] Thus sometimes we can have competing reference classes that in fact don't compete, but can be converted into estimates of consequents and thus consistently integrated.[22]

So you need follow only two basic rules for finding your initial prior probability: (1) use the narrowest, most clearly definable reference class whose contents are best known, and draw an *a fortiori* frequency from the data in that class; (2) and when you have equally competing classes and don't know what's in their conjunction, and you can't move one of them into e or don't know how, then use the conjunction of those two classes anyway by deriving a frequency from its *hypothetical* contents, as best you can estimate those contents to be. As for example in the Seguntium case: your overall background knowledge establishes that the conjunction set ('in Italy' and 'no special patronage'), though its contents are unknown, far more probably has contents similar to the 'no patronage' set than the 'in Italy' set—because you can present evidence proving that this is far more likely. The degree to which you are still uncertain as to these sets' nearness or agreement in this respect will then be reflected by the degree to which you expand your margins of error. As again in the Seguntium case: if you aren't sure the conjunction set has exactly a frequency of 0.20 ("one in five such colonies have public libraries"), you might still have sufficient evidence to be sure that that frequency can't be less than 0.10 ("one in ten such colonies have public libraries") or more than 0.33 ("one in three such colonies have public libraries"). And from there your analysis can begin.

Another question that comes up is how we can draw a hypothesis out of a reference class. If we are talking about causal hypotheses (such as which h best explains the bizarre content of Paul's letters), then the reference class will be a collection of the most relevant cause-effect pairs: letters with those features, paired with their causes (where $x \rightarrow y = x$ caused y), will generate a reference class in which some members (of quantity A) correspond to the cause-effect pairing A*{subject didn't exist}$\rightarrow${letters with those features} and other members (of quantity B) correspond to the cause-effect pairing B*{subject did exist + other causes}$\rightarrow${letters with those features}, and the ratio (A) / (A+B) is the prior probability that our event is a member of A* (or near enough). Scientists do this frequently, explicitly and implicitly, sifting through sets of cause-effect pairs to identify which causes typically explain an observed effect. If "seeing Venus" causes 20% of cases of 'UFO reported,' then the prior probability that a UFO report is simply an uninformed observation of Venus is 20%. But we usually have more information than that, both for the report and the class. For instance, if in cases of 'UFO reported with features {A, B, C}' 9 out of 10 times *that* report was caused by an uninformed sighting of Venus, then the prior probability that another such report was caused by seeing Venus is 90%. Hence causal reasoning is useful and commonplace.

It's also important to remember that prior probability is *not* the final probability. Admitting that a hypothesis has a small prior probability does not mean you concede that the hypothesis is unlikely to be true. The evidence for the specific case at hand must still be examined, and that evidence could confirm the hypothesis even when it has a low prior. Hence prior probability is not an assessment of the likelihood your hypothesis is true in the specific case you are examining. It's an assessment of how often such hypotheses are *ever* true, as a general rule—because it's logically necessary that prior probability judgments apply equally to all members of the same reference class. For example, if Jesus and Romulus belong to the same reference class (I'm not saying here that they do, only *if* they do), then the prior probability that a supernatural agency raised them from the dead must be the same (as both were reported to have been thus raised). Only if Jesus or Romulus can be shown to belong to a *relevantly* different reference class entailing a different prior probability for each of them does this rule of equality not obtain between them. But even then it still obtains between all members of their new respective reference classes (everyone Jesus is being paired with that distinguishes him from Romulus, or vice versa). And it's not enough to find just any different class, because every claim will belong to many different reference classes, and the rule of greater knowledge requires you to prefer the narrowest applicable class. Hence you can only derive a different prior if your hypothesis belongs to a different reference class that both satisfies the requirement of greater knowledge and entails a different prior probability. Formally, the hypotheses here are "Jesus was raised from the dead by a supernatural agency" and "Romulus was raised from the dead by a supernatural agency," and the proposed reference class they share in common is 'persons claimed to have been raised by a supernatural agency.' I have a hunch it's unlikely we'll ever find a narrower class that entails any difference in the priors and still satisfies the rule of greater knowledge (especially as that rule means, again, actual *knowledge*, not faith, belief, dogma, speculation, or assumption).

Just imagine that the actual evidence for your particular case is hidden behind a curtain, so you can't yet see it (so you don't know if you are looking at your case or any other case in the same reference class): how likely would you say it is that an $h \rightarrow e$ relation obtains in that case? In other words, assigning priors requires objectivity. For example, the prior probability that Jesus was raised from the dead by a supernatural agency is the same as the prior probability that a supernatural agency raised Romulus from the dead, or Asclepius, or Zalmoxis, or Inanna, or Lazarus, or the "many Saints" of Matthew 27:52–53, or "the Moabite" of 2 Kings 13:20–

21, and so on.[23] Obviously, the evidence in the case of Jesus can be much stronger than for Asclepius or any of these others, but that is accounted for with the consequent probabilities. Here we're only talking about the *prior* probability. You could again argue for a narrower reference class, but only if you have established data, for example, if Jehovah-miracles were disproportionately more confirmed as true than others, then Jesus belonging to the former class would allow us to assign a higher prior. But there is no adequate evidence to confirm that miracles associated with Jehovah are any more frequently true than others, nor to determine the frequency of true miracles among all Jehovah-associated miracle claims (since to this day we have yet to verify even one of them as genuinely miraculous, or in some cases as even having happened). See my discussion of the Smell Test in chapter 4 (page 114). All the same principles apply to ordinary cases, too.

Another issue historians must attend to is how reference classes change with context. For example, if a pie is stolen from a windowsill, the prior probability for any hypothesis about what happened to it will depend on the reference class 'pies stolen from windowsills,' but the specific contents of that class (and thus the frequencies that entail the priors) will depend on the context (which establishes a narrower reference class). If this happened at a cabin in the woods, the most frequent cause will be a raccoon; in a city, a person. And in a hypothetical scenario in which the world is filled with robots programmed to steal pies, the most frequent cause would be a robot. Thus when dealing with different cultures, eras, places, and social contexts, the content of any relevant reference classes will change accordingly. Which is again why specialized expertise is so important to doing history well (as argued in chapter 2).

Last but not least is the importance of dividing the 'probability space' among all possible explanations of the same evidence. To explain why we must do this and what this means, I'll conclude by introducing and analyzing a toy example—by which I mean an example that is invented solely to play with, in order to illustrate the logical principles underlying more real-world applications. Bayes's Theorem is like any scientific theory in this respect. Take, for example, the laws of motion. They must be formulated for ideal cases that don't exist in the real world. They then can be applied through various approximations to actual cases.[24] So, too, for BT: I will demonstrate how the model works using an imaginary perfect case, which never really happens, and then show how the model can then be approximated to apply to imperfect cases.

Let's start with the datum, which goes in e, that 'Joe is rich.' Assume this is securely in evidence and not in dispute. Let h = the hypothesis that "Joe won the California State Lottery (CSL)." Assume the established background knowledge is that Joe lives in modern California (and all else that entails). The prior probability of h then equals the number of people in Joe's circumstances ('living in modern California') who won the CSL, in ratio to everyone there, which includes people who didn't get rich (whether by winning the CSL or not) and people who got rich in some other way: theft, inheritance, successful business, stock market, drug dealing, finding a leprechaun's gold, gift from space aliens, etc., in other words, every logically possible thing, which all have some nonzero probability of being true and therefore must share some proportion of the total probability space for the reference class 'people in modern California.' A diligent study of modern records would allow a careful sociologist or historian to develop a credible estimate of the actual ratio of people in modern California who won the CSL.

Hence we are not talking about the mere probability of some person (like Joe) winning the lottery (which may be a million to one against), nor the probability of just anyone winning the lottery in any given population (which is often nearly 100%, because usually *someone* wins the lottery), since neither will produce any relevant probability that would correlate with ~h, and P(h|b) and P(~h|b) must always sum to 1, which means the prior probability of ~h must equal the sum of the prior probabilities of every other h (every other way someone in MC can get rich or fail to get rich). Thus, the population of MC can be divided among all possibilities (the number of lottery winners, business owners, day laborers, unemployed, and so on), and all those ratios (each one corresponding to a separate hypothesis explaining how Joe got rich) will sum to 1. This means you have to ignore the evidence of 'being rich' (vs. 'not being rich') and just throw every and any hypothesis into h. But hypotheses that have no chance of explaining e (e.g., "Joe has only ever been a grocery clerk") will have consequent probabilities near zero. And this allows you to ignore them, given suitable adjustments to the math.[25] The only way to ensure the sum always comes out to 1 is if you find the probability that Joe would win a lottery *relative* to all other possible ways of explaining e (in this case, that 'Joe is rich'), which is not the odds of winning a lottery, but the number of people in the reference class ('people in modern California') who won the CSL. There is no other way to ensure that $P(h|b)$ and $P(\sim h|b)$ will sum to 1. And since you do not usually know all possible alternative hypotheses (much less take them all into account when estimating prior probabilities), you have to work with the frequencies you have data for and leave the remaining probability space for 'all other possible causes of e.'

We would be able to do the same for any other h, though to a lesser degree in some cases. Let's suppose we test h_1 = the hypothesis that "Joe is a drug dealer" and then h_2 = "Joe found a leprechaun's gold" and h_3 = "Joe was given a ton of precious metals by space aliens." Though we know (from the contents of b available to us) that there are people who have gotten rich via h_1, and we know there are drug dealers in modern California (MC), we probably do not know the actual ratio, since by the nature of the illegal drug trade it's kept secret and to some extent off the radar of public records. However, if we engaged a sufficiently detailed investigation, we could develop relatively exact figures on how many people have gotten rich in MC (and for most of them, how) and how many haven't (and why), and that would leave a certain number unaccounted for, those for whom we could not ascertain these facts. True, we would have to answer the problem of defective records (e.g., someone lying on her tax forms about how she got rich or even whether she had), but this could also be overcome to within some margin of error with an FBI-scale investigation, or even any adequate investigation that takes account of all pertinent data (such as regarding the frequency with which such biases are known to exist in the available records). One way or another we can develop a defensible ballpark figure for the largest number of people who could possibly be making a living dealing drugs in MC, and we could use that as our prior probability. Or we can run the math with that as an upper bound and a lower figure—the *least* number of people who could be making a living dealing drugs in MC—as a lower bound. Or even widen that range to develop a conclusion *a fortiori*.

At every turn my estimates could be wrong (bad data, deception, errors in math, etc.), but Bayesian analysis would allow me to correct my calculations as new information is acquired (e.g., a critic could correct me by presenting actual data from which a different ratio can reasonably be inferred). My estimates would constitute testable predictions, and thus could be verified, falsified, or refined, with increasingly accurate information gathering. And as long as I specify my estimates as the number n 'who got rich or failed to get rich' by each method m, all the ratios I thus generate will sum to one (since adding up the n for each m will get me the total n of all people in MC, which is our narrowest available reference class).

All that is fairly straightforward. But now, what would the prior probability of h_3 be? Unlike drug dealing, nowhere in my background knowledge b is there any knowledge of there even being space aliens contacting MC at all, much less space aliens doing this *and* making people rich with gratuitous gifts. Though some of the people in MC *might* have gotten rich this way—some among the small set of those whose cause of wealth could not be determined by our inquiry and/or some among the rest, who otherwise lied about the source of their wealth—I have no reason to believe any have. Certainly, given my b (the b available to me), the ratio of such people to all other people in MC must be extremely small—for even if there are such people, they must be rarer even than lottery winners, unless there is an extremely impressive conspiracy going on worthy of a Philip K. Dick novel, and surely the prior probability of *that* is less even than the prior probability of anyone getting rich off of space aliens to begin with (per the effect on prior probability of stacking up *ad hoc* elements to explain away evidence as discussed in chapter 3, page 80). Therefore, though I do not have any scientific or even economic data on the 'number of people who got rich via space aliens' and even though I cannot, even with an FBI-scale investigation, eliminate such people from the set of all people in MC (known and unknown), I will still be acting rationally if, given my b, I assigned this a minimal prior probability—something significantly less than the prior probability of the first h, which I'll call h_0 ("Joe got rich by winning the CSL"). For surely, given my background knowledge, if there *are* such people, it would be objectively reasonable for me to believe they must be extremely rare (surely fewer than one in a million—and since zero *is* fewer than that, I can still say it's fewer than some nonzero number and consistently believe the actual number is indeed zero).

But now how about h_2? Unlike h_3, where there is at least the barest possibility from the information in b (as I do believe it is physically *possible* that there are space aliens and they could, if historical contingencies played out just right, have reached earth and given someone here a ton of gold—and those historical contingencies *do* have a real nonzero probability, however absurdly small it may be), there is not even the barest possibility from any information in b of there being leprechauns who hide their gold. But still, we can always be in error. So there is at least some nonzero *epistemic* probability that h_2 is true, so I must assign it a prior probability (if I want to be absurdly thorough). And I have reason to make it even less than h_3, since though both h_2 and h_3 must be *at least* as rare, and extremely rare at that (being equally not in evidence), I nevertheless have *some* reason in my background evidence to believe h_3 more likely than h_2. Of course, things could change, facts could develop that would persuade me to believe that there are more people in MC who got rich via h_2 than via h_3, and if so I would adjust my prior probabilities accordingly. But until then, I shouldn't.

If I somehow acquired rock-solid data on everyone in MC, so that I could be absolutely sure (i.e., certain to an absurdly high probability) no one had gotten rich in MC via h_2 or h_3, I might want to set their prior probability at zero. But I still couldn't, because there is still some chance that my data is in error somehow, that I missed someone, or that Joe is the first instance (and thus all the prior cases could not have entailed a probability of zero anyway), and so on.[26] However, the probability that I am in error creates an upper bound to the probability that someone got rich via h_2 or h_3—that is, the certainty that I have (that no one did) directly limits the epistemic probability (at least for me) that someone, nevertheless, did (a point I'll revisit on page 253). Hence the prior probability can't be zero but it must be extremely low. Obviously I can never know with scientific certainty what that prior probability is (and in actual physical fact, it could well be zero, but we can never know that *for sure*). All I can do is work with what is reasonable to me given my *b*, and that's what BT entails I should do. And in the end, as I've explained in chapter 3 (page 70), when we are dealing with hypotheses with priors this astronomically low, we can actually just treat them as having a prior probability of zero, since it will make no difference to the outcome—*unless* we have specific evidence in this case that it might yet be true. Thus, while the probability space must be divided among infinitely many hypotheses (all logically possible hypotheses that explain *e*), we can ignore all but a finite few of them, and simply divide that space among those.[27]

Now what happens when we change historical context? Suppose we are testing *h* = "Matthias the first-century Galilean got rich by building industrial machinery." Unlike MC, the circumstances (here we'll posit early-first-century Palestine) do not admit of any comparable data. We have only a vague idea what the ratios were of means-to-wealth in that period and place. Moreover, the ratios were very different then than in MC. For example, building industrial machinery is a common career in MC, but much less so in ancient Galilee. It was not, however, nonexistent, as by then the Romans had disseminated a variety of water- and animal-powered machines for industrial functions as diverse as processing flour to sawing stone, and someone had to build them (and having a rare skill that was in demand among the ancient equivalent of big business, they likely made good money at it). But this only makes the problem harder, since we can't rule them out (we know such persons probably existed), yet their relative numbers must have been very small. This leaves us with two options: conclude that the matter is hopeless and beyond any further analysis, or go forward with a Bayesian analysis using what little we do know and *a fortiori* estimates.

All we have in this case are a hoard of scattered out-of-context tax records from Egypt that present very limited information, a vast body of useful but often ambiguous archaeological and epigraphic data, and a large body of literary evidence regarding rich people and how they got rich, which is highly subjective and incomplete. But that's all we have, so we either give up and say BT is useless here, or we go forward and ask what BT compels us to conclude given what we *do* know. Since BT underlies all valid historical methods, concluding that BT can't solve the problem entails concluding that no method can, and that therefore we can never assert anything about whether the report that "Matthias the Galilean got rich by building industrial machinery" is true or false. Such radical skepticism is excessive and unwarranted. We can certainly say *something* about that statement's likelihood of being true, for example, we can surely say it's far more likely to be true than "Matthias the Galilean got rich by transmuting lead to gold" or even "Matthias the Galilean got rich by winning the California State Lottery" (or any lottery, since so far as we know, none existed then).

In other words, it should be reasonable to say that, given my considerable expertise in the relevant culture and economy (having actually translated Egyptian tax receipts, having actually read widely in the field of Roman economics, having read a considerable body of diverse literature from the time, having studied a diverse array of inscriptions, etc.), I can honestly say that probably most people in early-first-century Galilee got rich by inheritance, then after that most of the remainder of those who got rich did so through gifts, graft, or bribery, and then after that most of the remainder got rich through a plethora of occupations and business ventures (and again, I'd change these relative proportions in light of any sufficient evidence to the contrary). And perhaps after that we'd have riches procured by brute crime (such as robbery and piracy), by which point we are at very small numbers—and even smaller still would be, again, "finding a leprechaun's gold" and "given a pile of gold bars by space aliens," since we have no information in *b* that would suggest the frequencies of these things have changed at all between then and now, or that they were ever as frequent as, say, successful piracy.

Putting all this together, logic compels a conclusion. Here we are taking as a given that Matthias was rich and only asking whether industrial mechanics is how indeed he got that way, taking as evidence anything we can find attesting to that. I know enough to say that the prior probability that anyone (and hence our hypothetical Matthias) got rich by any occupation or business venture at all (much less that specific one) is probably significantly less than 0.2. This is because the data available to me justify the reasonable conclusion that by far most rich people in ancient Galilee (AG) got that way by inheritance, and that *must* equate to *at least* 0.6, leaving 0.4 left over, and since the data also justify the reasonable conclusion that most of the remaining rich people in AG got that way via graft and other forms of corruption or benefaction, that leaves no more than 0.2 unaccounted for. If we rule all other possibilities (e.g., a remarkably successful career in burglary, stumbling onto a gold mine by accident, leprechauns, aliens, etc.) as extremely rare (let's say, altogether, less than one in ten million) we can disregard them as negligible with respect to h and stick to our maximum possible prior probability of 0.2. But that was for any business venture at all, and h proposes a very specific and in fact uncommon business venture (industrial mechanics). So the prior probability of getting rich *that* way must not only be less than 0.2, but substantially less. I would say at least less than a hundredth of that (there were more than 100 possible occupations and wealth-making business ventures in antiquity and most were more common than this one), which leaves us with less than 0.002; but I also have to say it could possibly not have been less than one in a million (as there cannot have been more than one million rich people in ancient Galilee, but given the nature of the economy and other facts in b, there had to have been at least one wealthy industrial mechanic), which leaves us with a range of priors of 0.000001 to 0.002. Depending on our aim, we can use either as our *a fortiori* estimate or calculate what exact frequency within that range would be required to believe Matthias got rich that way given what evidence we have. Either conclusion can be of use to a historian, depending on her objectives.

Such low priors should not alarm anyone. We can easily have in e items of evidence that would be equally improbable unless h were true. With sufficient good evidence, even a prior of 0.000001 can leave us with a posterior probability near enough to 100% to fully believe the claim. If we had a contemporary historian describing Matthias's enterprises and success, depending on the content of that account we might be able to say that the probability that a historian would report such a story if it were false is substantially less than if it were true—enough, in fact, to conclude that more probably than not, the story is true. Factors that would convince me of this include: the very idea that someone could get rich that way suggests (at least more than chance) that it derives from some actual experience rather than fantasy; the specific details reported of how it was done, if they all track actual and otherwise obscure facts of the time (as that is much less likely for a fabrication); and the absence of any likely or discernible reason for the story as we have it to have been made up (by the historian or his source). These all constitute elements of e, and collectively they would be extremely improbable unless the story is true. That might be debatable—but the fruits of such a debate, with its inevitable focus on specific and comparative evidence and its logical significance, is precisely what I find of use in approaching such questions with BT.

In contrast, of course, if we were to propose "Matthias got rich by transmuting lead to gold," the prior probability of *that*, given b, would have to be *extraordinarily* small, much less than any historian or his sources is likely to be trustworthy, and therefore it's very unlikely any historical evidence can convince us. We'd have to confirm independently that it was even possible first (such as actually transmuting lead to gold in a laboratory using technologies Matthias would have had access to), which discovery admittedly would greatly increase the prior probability (since for a number of sound reasons we assume physics that works now, worked then). But unlike transmuting metals, even without a single confirmed case we can conclude the frequency of industrial mechanics getting rich in AG was not zero, and in fact much more likely (vastly more likely) than transmuting lead to gold, because all the factors required for the former (in terms of culture, physics, economics, etc.) can be independently confirmed to have existed then as now, and all we are querying is the probability of their conjunction. But in the case of transmuting lead, we have not even *any* element confirmed (and in fact several contradicted) in extant evidence, and thus it's not a question of the conjunction of known causes, but the existence of entirely *unknown* causes, which if they existed at all, should have been discovered by now.[28] Thus, very small priors should not be confused with *absurdly* small priors.

And once again we must not forget that the prior probabilities are relative probabilities: the total probability space is divided among all hypotheses, because all priors must sum to one. Hence, for example, in a family poker game the prior probability of being dealt a royal flush by natural accident is *not* the probability of being dealt a royal flush (which is extremely small), but the probability that when royal flushes are dealt, they are dealt by natural accident rather than something else, e.g., cheating or a bad deck. Which depends on the reference class—in other words, the context. This should be obvious, because, after all, once we assume either happened ('natural accident' or 'something else'), the consequent probability on both is normally the same (whether it was fair or a cheat, you will observe exactly the same result: a royal flush), yet we routinely assume natural accident when we get an awesome hand at cards (especially in a family game), therefore we obviously must believe the prior probability for 'natural accident' is *very high*, and we would demand good evidence of it being otherwise before believing it was (i.e., evidence that heavily skewed the consequents toward some alternative explanation). This reflects the fact that we are relying on past experience regarding the dealing of poker hands, and our background knowledge regarding the skills, technologies, circumstances, and motives required to rig such a hand. In short, in such contexts (e.g., a family game), we find the frequency of cheating (and bad decks and so on) is very low (even collectively, much less individually) compared to natural and fair deals. Hence with a reference class of 'amazing poker hands dealt' (or more specifically 'amazing poker hands dealt in a family game,' although other contexts are likely to make little difference, e.g., we wouldn't normally expect cheating or error even at a casino) we estimate the relative frequency of the different causes of an event being in that reference class (the event in this case being 'an extremely convenient and unlikely draw of cards'). Which is still, of course, a frequency, but not the frequency of drawing a royal flush by chance, but the frequency of drawing a royal flush as a result of chance *relative* to the frequency of drawing it as a result of some other cause (all conditional, of course, on *b*, which in this case includes facts such as what a "family game" means and all our past experience and knowledge mentioned previously). In other words, if the reference class 'all draws of a royal flush in a family game' contained 998 fair draws and 2 cheats, the frequency of fairly drawn royal flushes is actually extremely high (0.998). Hence, in that context, the prior probability that a royal flush was fairly drawn is 99.8% (or near enough).[29]

So much for prior probabilities. Assigning consequents follows similar rules, only the consequents don't have to sum to one. The probability space for priors consists of all members of the most pertinent known reference class. But the probability space for consequents consists of what the hypothesis predicts; more specifically, it consists of all logically possible consequences of h that are mutually exclusive of each other (of which e must be one such, for h to have a nonzero probability of causing it, and thus for the consequent to be anything but zero). Again, "logically possible" includes even the absurd, but the absurd takes up such a minuscule proportion of the probability space it makes no practical difference and can usually be ignored. For example, the consequent probability that a medieval alchemical procedure will transmute lead to gold is vanishingly small; therefore the consequent probability that it will have some other effect far more mundane is as near to 100% as makes all odds (unless we have information in b that suggests the procedure's success is a credible possibility). Hence consequent probabilities can be derived using this same subtractive reasoning. For example, if we said $P(e|h.b) = 0.8$ where h = "Joe won the CSL" and e = "Joe is rich," then *what we are saying* is that 1 out of 5 times when someone wins the lottery they do not get rich (and hence there is a 20% chance of that outcome), and the other 4 out of 5 times they do get rich (and hence that has an 80% probability). Otherwise, if we believed winning the lottery was virtually 100% certain to have this result (if everyone who wins gets rich, and the number of confirming cases is in the hundreds or even thousands), we would have to believe that, practically speaking, given h, then always e, which we would instead represent with $P(e|h.b) \approx 1$. Or if the information we have in b entails that 80% of those who win lotteries *don't* get rich, then we would be compelled to believe that $P(e|h.b) = 0.2$. In other words, $P(e|h.b) = 1 - P(\sim e|h.b)$, which is a useful rule to have at hand, since determining $P(e|h.b)$ is often made easier by determining $P(\sim e|h.b)$ instead, and then taking the converse of that. And if following this procedure gets you a different result for P(e|h.b) than you previously did, then you know you've done something wrong. Thus, "How often would the hypothesized facts *not* produce (or not be correlated with) *e*?" is always a question well worth asking.

I suspect many critics by now have been chomping at the bit in protest of my cavalier assumption that epistemic probabilities in Bayes's Theorem are really just actual frequencies of things, and thus really physical probabilities after all—since I have evinced this assumption throughout this chapter and most of this book. I've given some hints already as to why that assumption is in fact valid, and as to how we convert such frequencies into seemingly unrelated things like "degrees of belief," but I'll take that up in the rest of this chapter (especially in the last section, to which the next section is preliminary). In closing here, one final point deserves mention: even if we err in choosing a reference class (for priors) or estimating causal frequency (for consequents), this is no different than any other error in empirical reasoning. Until the error is identified and corrected, or shown to be uncorrectable, we have sufficient reason to believe what our analysis tells us. And we can be corrected, and will thus change our minds, if that error is indeed exposed by critics and then corrected, either by them or our own renewed inquiry and analysis.

If we can legitimately narrow the reference class, or are compelled by the logic of the situation to broaden it, we would simply recalculate our conclusion accordingly once we've been given new information not previously available to us. So, too, the solution to all other difficulties that arise in applying BT. And as we are discussing conclusions in history and not science, per my discussion in chapter 3 of the difference in degree between those two enterprises (page 45), as long as you follow all the rules and advice above and throughout this book, a Bayesian analysis will show you what you should believe *given what you know*, and since most of what you know (most of what's in *b* and *e*) does not rest on scientific certainty, neither can any conclusion you reach via BT. But scientific certainty is not required to warrant ordinary belief. What *is* required is that whatever degree of certainty you settle upon, it be based on a well-informed and logically valid analysis.

THE ROLE OF HYPOTHETICAL DATA
IN DETERMINING PROBABILITY

What are probabilities really probabilities *of*? Mathematicians and philosophers have long debated the question. Suppose we have a die with four
sides (a tetrahedron), its geometry is perfect, and we toss it in a perfectly
randomizing way. From the stated facts we can predict that it has a 1 in
4 chance of coming up a '4' based on the geometry of the die, the laws
of physics, and the previously "proven" randomizing effects of the way
it will be tossed (and where). This could even be demonstrated with a
deductive syllogism (such that from the stated premises, the conclusion
necessarily follows). Yet this is still a physical probability. So in principle
we can connect logical truths with empirical truths. The difference is that
empirically we don't always know what all the premises are, or when or
whether they apply (e.g., no die's geometry is ever perfect; we don't know
if the die-thrower may have arranged a scheme to cheat; and countless
other things we might never think of).[30] That's why we can't prove facts
from the armchair.

Nevertheless, Archimedes was able to prove the existence and operation of mathematical laws of physics purely from deductive logic—and he
was right (he thus derived the basic laws of leverage and buoyancy). But
he was only right because the premises on which his syllogisms depended
were empirically confirmed to his satisfaction—at least in those conditions
he restricted his laws to. We now know there are many factors that can alter
or negate his premises and, therefore, to be more broadly applicable, his
laws had to be considerably revised and expanded.[31] He could not have
deduced the world would turn out that way. But he could have speculated
it would and then correctly deduced what laws of physics would then
follow. And he was aware of this. For example, he knew the curvature of
the earth complicated his premises for determining the laws of hydrostatics
(since it meant the surface of a tub of water would not be a flat plane but
a rounded convex shape), so he proved that that curvature was so slight it
could be safely ignored for his purposes (i.e., he could assume the surface
of a tub of water is flat). Only if someone demanded a certain (he might

say absurd) level of precision would that curvature have to be reintroduced and accounted for. The point being, empirical facts are not so different from logical "facts"; it's just that when the information available to us is inescapably limited, we can no longer use logic to ascertain the facts. We must still be logical, but we have to go and look at the world and collect the information we need, which is always limited in such ways that we can never be sure that what seems to be the case really is the case. Geocentrism is a famous example: the universe seems to rotate around the earth. In this case, how things seem, it turns out, is not how they actually are. More information eventually revealed that to us. And it's conceivable (albeit absurdly improbable) that yet more information could reveal we were wrong the second time and should have stuck with the way things seemed (e.g., maybe space aliens using ultra-advanced technology are altering the empirical data to fool us into thinking the solar system is heliocentric when it's actually geocentric). The only way to know is to gather more information. And when we don't have it, we can only say what's likely to be true given what we know at that moment.[32]

Thus we go from logical truths to empirical truths. But we have to go even further, from empirical truths to hypothetical truths. The frequency with which that four-sided die turns up a '4' can be deduced logically when the premises can all be ascertained to be true, or near enough that the deviations don't matter (like the curve of the earth for Archimedes), yet "ascertained" still means empirically, which means adducing a hypothesis and testing it against the evidence, admitting all the while that no test can leave us absolutely certain. And when these premises can't be thus ascertained, all we have left is to just empirically test the die: roll it a bunch of times and *see* what the frequency of rolling '4' is. Yet that method is actually less accurate. We can prove mathematically that because of random fluctuations the observed frequency usually won't reflect the actual probability. For example, if we roll the die four times and it comes up '4' every time, we cannot conclude the probability that this die will roll a '4' on the next toss is 100% (or even 71%, which is roughly the probability that can be deduced if we don't assume the other facts in evidence).[33] That's because if the probability really is 1 in 4, then there is roughly a 4% chance you'll see a straight run of four '4's (mathematically: $0.25^4 = 0.00390625$). Since you don't know which situation you are in (the latter one, in which the die rolls fairly, or the former one, in which it always lands on '4'), you're stuck. You thus can't deduce from the observed data alone what the odds of the next toss really are. You *can* deduce, however, that with a *lot* more data you'll be on safe enough ground to do that. If you roll the die a thousand times and it still comes up '4' every single time, you've pretty well proven the die is rigged. Though there is a calculable chance that this is a fluke and that the odds of any roll being a '4' are and always have been 1 in 4, that probability is so incredibly small you will be quite safe in assuming that's not what's going on (in fact that probability is 0.25^{1000}, which is unimaginably small). It's possible to work out mathematically when and why (and how much) we should trust a given conclusion from a given set of data, all just from a comparison of the infinite possibilities and their probabilities.[34] But you can always still be wrong. All we can do is calculate the best bet.

The point of the above example is that we can't simply rely on actual data sets.[35] Even a thousand tosses of an absolutely perfect four-sided die will not generate a perfect count of 250 '4's (except but rarely). The equivalent of absolutely perfect randomizers do exist in quantum mechanics. An experiment involving an electron apparatus could be constructed by a competent physicist that gave a perfect 1 in 4 decision every time. Yet even *that* would not always generate 250 hits every 1,000 runs. Random variation will frequently tilt the results slightly one way or another. Thus, you cannot derive the actual frequency from the data alone. For example, using the hypothetical electron experiment, we might get 256 hits after 1,000 runs. Yet we would be wrong if we concluded the probability of getting a hit the next time around was 0.256. That probability would still be 0.250. We could show this by running the experiment several times again. Not only would we get a different result on some of those new runs (thus proving the first result should not have been so concretely trusted), but when we combined all these data sets, odds are the result would converge even more closely on 0.250. In fact you can graph this like an 'approach vector' over many experiments and see an inevitable curve, whose shape can be quantified by mathematical calculus, which deductively entails that that curve ends (when extended out to infinity) right at 0.250. Calculus was invented for exactly those kinds of tasks, summing up an infinite number of cases, and defining a curve that can be iterated indefinitely, so we can "see where it goes" without actually having to draw it (and thus we can count up infinite sums in finite time).

Clearly, from established theory, when working with the imagined quantum tabletop experiment we should conclude the frequency of hits is 0.25, even though we will almost never have an actual data set that exhibits exactly that frequency. Hence we must conclude that that hypothetical frequency is more accurate than any actual frequency will be. After all, either the true frequency is the observed frequency or the hypothesized frequency; due to the deductive logic of random variation you know the observed frequency is almost never exactly the true frequency (the probability that it is is always ≤ 0.5, and in fact approaches 0 as the odds deviate from even and the number of runs increases); given any well-founded hypothesis you will know the probability that the hypothesized frequency is the true frequency is > 0.5 (and often $\gg 0.5$, and certainly not $\to 0$); therefore P(THE HYPOTHESIZED FREQUENCY IS THE TRUE FREQUENCY) $>$ P(THE OBSERVED FREQUENCY IS THE TRUE FREQUENCY); in fact, quite often P(HYPOTHESIZED) $\gg$ P(OBSERVED). So the same is true in every case, including the four-sided die, and anything else we are measuring the frequency of. Deductive argument from empirically established premises thus produces more accurate estimates of probability.

Technically, sure, this would be moot when we have perfect data, for example, if we were 100% certain our data set included, without error, every single cause-effect pair in our reference class that has ever existed and ever will, *even including the case we are testing*, then the observed frequency, not the hypothesized frequency, is the prior probability. But we never have perfect data. Nor can the case we are testing already be fully described in such a reference class, as then we would already know whether h was true. Nor can our case be the only other case, for at the very least the logical possibility remains that there will be many more, and in most cases we would deem this quite probable, and even when we don't, every logical possibility still entails some nonzero epistemic probability. Thus our prior must reflect the probability that our one new case (as well as any other cases not yet collected) will conform to the frequency exhibited by all other cases (known and unknown), and that requires application of a hypothetical frequency, just as with the die roll. For instance, if a four-sided die were rolled only ever four times in the whole of human history up to now, and it came up four '4's, the observed frequency would give us a wildly incorrect prior probability that the next roll will be a '4.'

This conclusion, that hypothetical frequencies are more accurate than observed frequencies, should not surprise anyone. For the same is true of most human knowledge. For example, we do not conclude the universe consists of only what we see at any one moment, but that it consists of what we hypothesize exists behind walls and obstacles and everywhere else our vision does not directly penetrate. And as a result, we much more success-fully navigate the world. Thus, again, hypothesis beats direct observation. In just the same way, if we take care to manufacture a very good four-sided die and take pains to use methods of tossing it that have been proven to randomize well, we don't need to roll it even once to know that the hypo-thetical frequency of this die rolling '4's is as near to 0.25 as we need it to be. Even more so if we use *a fortiori* estimates instead. For example, if we said the odds of rolling a '4' with that die are 'at least' 0.20, we'll be even more certain that that's true than if we declared it to be "exactly" 0.250. Because the probability of being wrong in the former case is vastly smaller. And the further out we go, the smaller that probability gets; for example, if we said "at least" 0.10, then the probability of our being wrong will be vastly smaller still.

Thus it's not valid to argue that because hypothetical frequencies are not actual data, and since all we have are actual data, we should only derive our frequencies from the latter. All probability estimates (even of the very fuzzy kind historians must make, such as occasioned in chapters 3 through 5) are attempted approximations of the true frequencies (as I'll further explain in the next and last section of this chapter, starting on page 265). So that's what we're doing when we subjectively assign probabilities, attempting to predict and thus approximate the true frequencies, which we can only *approximate* from the finite data available—because those data do not reflect the true frequency of anything (but rarely, and we'll never know when, so the fact that sometimes they will is of no use knowing). Thus we must instead rely on hypothetical frequencies, that is, frequencies that are generated by hypothesis using the data available—which data includes not just the frequency data (from which we can project an observed trend to a limit of infinite runs), but also the physical data regarding the system that's generating that frequency (like the shape and weight distribution of a die). Of course, when we have a lot of good data, the observed and hypothetical frequencies will usually be close enough as to make no difference. But it's precisely because historians rarely have such good data that they must know how to construct hypothetical frequencies from the data they do have.

Indeed, some historical events literally happen only once, because the conditions required converged only once. Yet we need to know with what frequency such a conjunction of causes will produce that effect. Returning to our previous example of Matthias the mechanic, suppose we had no evidence of anyone getting rich as an industrial mechanic in antiquity. That does not mean no one did. Because if it was rare, we can expect (to a very high probability) that we would have no evidence of it (see my earlier discussion of lost evidence, page 219). Meanwhile, all the elements required for it to have happened are well attested as operating in that context (and there is no evidence they were ever mutually exclusive), so their conjunction must have had an actual frequency. Indeed, this is so even if there were no rich mechanics. Just as a die that is never rolled nevertheless has a discernible probability of coming up '4' if it ever is rolled, so, too, the conjunction of conditions required to produce a rich mechanic in antiquity will have *some* probability even if, by chance, that conjunction never occurred. And when we actually are faced with evidence of such a conjunction (which is *always* a logical possibility), we are certainly required to assess the prior probability of that conjunction, even in the absence of any prior examples, because if we know anything, we know it can't be zero, and is unlikely to be vanishingly small (since, after all, we're not talking about transmuting lead to gold). And as it happens, in this case we *have* been faced with evidence of just such a conjunction, by discovering direct evidence of a wealthy Roman industrial mechanic who made his fortune in the Middle East.[36]

Even if we didn't have such evidence, anything we argued will still always presume some prior probability anyway, and it's better to have a well-considered one than one you've never questioned or even realized you were using. For example, if you reject the new evidence out of hand by arguing "there can't have been any such people back then," then you are implicitly saying the prior probability is virtually zero (or, in any case, too low for the evidence to overcome); or if you accept the new evidence as confirming "there were such people back then," then you are implicitly saying the prior probability is *not* virtually zero, but in fact high enough that this evidence is sufficient to make that claim very probable. Either way, you are assuming you know what that prior probability is (to some measure of precision).[37] How can we claim to know what it is? By constructing a hypothetical probability, just as we would do for a die we carefully made but never rolled. And that's exactly what you will have done (in either scenario, whether rejecting the evidence or accepting it) whether you realize it or not.

Thus, again, avoiding BT does not avoid constructing hypothetical frequencies; it just hides from view (and thus from criticism and test) the fact that you are depending on them. As historians, we routinely have to make hypothetical predictions of frequency based on models of what happened, which we construct from the facts at hand (our evidence *e* and our background knowledge *b*). So are we going to do that honestly and openly? Are we going to apply a valid and sound approach to that task? We certainly should. As long as we use *a fortiori* estimates (i.e., allowing large margins of error) and validly derive probabilities from our models and evidence, our predictions will be as accurate as any historical argument could ever make them.

Critics might still ask what we do with a true "single case" scenario, one in which we have no prior knowledge informing us (even by hypothesis) whether its existence or occurrence is any more likely than not. But that's actually the easiest case of all: if we have no relevant knowledge *at all* for determining its prior probability, then its prior probability is *by definition* 0.5 (I discussed this in chapter 3, page 83, and chapter 4, page 110). But such cases almost never occur in real life (except by stipulation, e.g., as when we begin a series of BT equations from scratch, by the conclusion of which we will have addressed all relevant data, but at the start of which we presume the complete absence of data—a procedure that is only valid because we do eventually introduce all known data before reaching the final conclusion: see chapter 5, page 168). We usually have many well-understood causal models for all the individual contents of any e, and though the conjunction of those elements may be unique, none of the individual elements are, and since we can often adduce probabilities for each element, their conjunction entails a probability of its own (barring none are mutually exclusive and we include among the required elements the absence of prohibiting or interfering causes). In fact, even when we can't adduce *any* probability for an element, that means *so far as we know*, its probability is 0.5; as otherwise, we would be able to adduce evidence that it's not 0.5 (as explained in chapters 3 and 4, pages 83 and 110). Since BT requires us to input what we know (because all probabilities in BT are conditional on background knowledge b) in order to generate conclusions that likewise represent what (at best) we know, it follows that we can always adduce probabilities for every possible cause of every element of any conjunction of elements constituting e. The only question is how accurately—which *a fortiori* estimates will answer, and new information can correct. (If you ask how an event can happen only once in all of history and yet not have a vanishingly small probability, the answer is that the probability changes over time; e.g., the probability that Churchill and FDR will win another world war is vanishingly small because Churchill and FDR no longer exist, so the probability of their doing anything now is vanishingly small, whereas the probability that Churchill and FDR would win World War II was not vanishingly small because they were then alive and in command of the Allied forces.)

None of the foregoing entails rejecting a frequentist interpretation of probability. To the contrary, probabilities are unintelligible on any other interpretation (a point I'll return to in the next section, page 265). That we must rely on hypothetical data sets only means that the frequency we are talking about when we talk about probabilities is the frequency we have the most reason to expect will obtain in any new or randomly isolated cases. And that frequency is the hypothetical frequency. Of all the frequencies we can expect to appear in a future sequence of rolls of our four-sided die, we have more reason to expect the frequency of 0.25 for each number than any other frequency. We'll be wrong (for any finite series of cases), but only marginally—and we'd be even more wrong if we expected any other frequency instead (in fact the probability of being wrong increases in direct proportion to how far our prediction deviates from the best-estimated hypothetical frequency). And the reason that's the case is that we have already confirmed (to a very high probability) a system of hypotheses about what that die is and how the world works, and so on, or else confirmed a system of hypotheses about the behavior of random processes, a collection of observed data, and a definable trend toward a limit—or both—and either system of hypotheses entails one expected frequency more than any other. But it's still a frequency we're talking about. Thus it's still a physical probability we're talking about.

BAYESIANISM AS EPISTEMIC FREQUENTISM

The debate between so-called 'frequentists' and 'Bayesians' can be summarized thus: frequentists describe probabilities as a measure of the frequency of occurrence of particular kinds of event within a given set of events, while Bayesians often describe probabilities as measuring degrees of belief or uncertainty.[38] But there really is no difference. That's what I'll set out to prove here.[39]

Probability is obviously a measure of frequency. If we say 20% of Americans smoke, we mean 1 in 5 Americans smoke, or in other words, if there are 300 million Americans, 60 million Americans smoke. When weathermen tell us there is a 20% chance of rain during the coming daylight hours, they mean either that it will rain over one-fifth of the region for which the prediction was made (i.e., if that region contains a thousand acres, rain will fall on a total of two hundred of those acres before nightfall) or that when comparing all past days for which the same meteorological indicators were present as are present for this current day we would find that rain occurred on one out of five of those days (i.e., if we find one hundred such days in the record books, twenty of them were days on which it rained).

Those are all physical probabilities. But what about epistemic prob-abilities? As it happens, those are physical probabilities, too. They just measure something else: the frequency with which beliefs are true. Hence all Bayesians are in fact frequentists (and as this book has suggested, all frequentists should be Bayesians). When Bayesians talk about probability as a degree of certainty that h is true, they are just talking about the fre-quency of a different thing than days of rain or number of smokers. They are talking about the frequency with which beliefs of a given type are true, where "of a given type" means "backed by the kind of evidence and data that produces those kinds of prior and consequent probabilities." For example, if I say I am 95% certain h is true, I am saying that of all the things I believe that I believe on the same strength and type of evidence as I have for h, 1 in 20 of those beliefs will nevertheless still be false (and h could be that one out of twenty—but since that's *de facto* unlikely, indeed the odds are 20 to 1 against it, I am warranted in believing h, at least as pro-visionally as a 95% certainly would allow). Probability can be expressed in fractions or percentile notation, but either is still a ratio, and all ratios by definition entail a relation between two values, and those values must be meaningful for a probability to be meaningful. For Bayesians, those two values are 'beliefs that are true' and 'all beliefs backed by a certain compa-rable quantity and quality of evidence,' which values I'll call T and Q. T is always a subset of Q, and Bayesians are always in effect saying that when we gather together and examine every belief in Q, we'll find that n number of them are T, giving us a ratio, n_t/n_q, which is the epistemic probability that any belief selected randomly from Q will be true.

The whole debate between frequentists and Bayesians, therefore, has merely been about what a probability is a frequency of, and that is a rather pointless disagreement, since a frequency is a frequency, the rules are the same for either, and therefore both sides are right, just about different things. For example, "the probability of rolling a '1' on a fair six-sided die is approximately 17%" is a statement of physical probability. It's the actual expected frequency, based on either the actual historical frequency or the hypothetical frequency given a certain set of confirmed physical condi-tions or data (see discussion in the previous section), but in either case a statement that can be unpacked the same way: if "the probability of rolling a '1' on a fair six-sided die is approximately 17%," then "a fair six-sided die rolled a very large number of times will roll a '1' approximately 17% of those times." In contrast, "there is a 17% chance that my belief is true that a fair six-sided die will roll a '1' on this next toss" is a statement of epistemic probability (entirely synonymous with "I am 17% confident that a fair six-sided die will roll a '1' on this next toss"), yet it's fully entailed by the statement of physical probability. Thus, when conjoined with certain analytic statements about belief, the latter becomes synonymous with the former.

This is because if the physical probability is confidently known to be 17% (as it is in that last example), there can be no rational basis for the *epistemic* probability to be anything else but 17% as well. The two only deviate insofar as data is lacking to know what the physical probability really is, but even then epistemic probability is still an estimate of physical probability, and is adjusted as those estimates are improved. Hence, epistemic probability is always an approximation of some physical probability. And as such, epistemic probability always approaches some corresponding physical probability as more data is acquired. Statements of epistemic probability thus amount to saying "if *these* physical probabilities measured in BT fall within *these* bounds, then the probability that any belief (that is logically entailed by those physical probabilities) will be true, will be *this* (i.e., what BT calculates)," or, put another way, "the number of beliefs (entailed by these physical probabilities) that will be true is *thus*," and therefore the number of beliefs, based on these physical probabilities, that will be false is the converse of that. Which of course entails we must necessarily have many false beliefs. But the more rationally and informedly we form our beliefs, the fewer of them that will be false, and if we proportion our belief to the evidence appropriately, we are thereby acknowledging the frequency of beliefs of a given type (i.e., beliefs based on a particular quality of evidence) that are false. So when you say you are only about 75% sure you'll win a particular hand of poker, you are saying that of all the beliefs you have that are based on the same physical probabilities available to you in this case, 1 in 4 of them will be false without your knowing it, and since this particular belief could be one of those four, you will act accordingly.

So when Bayesians argue that probabilities in BT represent estimates of personal confidence and *not* actual frequencies, they are simply wrong. Because an 'estimate of personal confidence' is still a frequency: the frequency with which beliefs based on that kind of evidence turn out to be true (or false). As Faris says of Jaynes (who in life was a prominent Bayesian), "Jaynes considers the frequency interpretation of probability as far too limiting. Instead, probability should be interpreted as an indication of a state of knowledge or strength of evidence or amount of information within the context of inductive reasoning."[40] But "an indication of a state of knowledge" *is* a frequency: the frequency with which beliefs in that state will actually be true, such that a 0.9 means 1 out of every 10 beliefs achieving that state of knowledge will actually be false (so of all the beliefs you have that are in that state, 1 in 10 are false, you just won't know which ones). This is true all the way down the line. To say "I am 99% confident that x will happen roughly 80% of the time" is to assert a confidence level, and a confidence level is a mathematically defined state (it follows necessarily from the deductive truths of randomized sets and some relevant physical frequencies), thus what you are saying is that the frequency of all beliefs in the same mathematically defined state that are true is 99 in 100. And of course the 80% in which you have this confidence is a straightforward physical frequency.

This is why every claim has some small prior probability. If you are 99% confident that $P(h|b) = 0$, that amounts to saying there is a 1% chance that $P(h|b) \neq 0$, so the *a fortiori* $P(h|b)$ must be something more than 0. And in fact, since it's never strictly correct to say we're 99% confident that $P(h|b) = 0$, but rather we must say that $P(h|b) = 0 +/-0.01$ (or whatever confidence interval we intend at that stated confidence level—see chapter 3, page 87), we must always accept the possibility that the actual value is at the upper end of that error margin (since we can only narrow it further by reducing our confidence level below 99%). Therefore, since we are never 100% confident,[41] nearly every logical possibility has a nonzero prior (hence my fourth axiom in chapter 2, page 23).

So perhaps we could say Jaynes is confusing confidence level with the frequencies in which we have confidence. Yet both are still just frequencies, confidence level being the frequency with which such confidence intervals correctly describe a physical frequency, and those confidence intervals describing a physical frequency (of some kind of event or correlation). If a frequentist validly determines that the frequency of x in a given population (in other words, a given reference class) has a 99% chance of falling somewhere between 20% and 30% (e.g., a frequency of 25% +/-5%), then a Bayesian must agree. That is, they must admit that their confidence that some new instance in that same reference class will have property x cannot be higher than 30% or lower than 20%, except 1% of the time, because (as the frequentist will have shown) the evidence can support no other conclusion. The former is their confidence interval; the latter, their confidence level. And when running a BT analysis in history, it's unwise to proceed with anything but a very high confidence level (well above 99%, high enough in fact that it shall never have to be stated), which we can always ensure by using *a fortiori* reasoning (as explained in chapter 3, page 85). Because the confidence levels for each probability assigned in a BT formula must commute to the conclusion. The mathematics of this can be quite complex, but as long as you keep your confidence levels very high, there will be no need to run any calculations—the confidence level that will always commute to the conclusion will then be the highest attainable for you (because you will only have chosen probabilities to include in your analysis that you can assert with the highest confidence you can attain). And that's all the certainty you need (or at least, all you can ever have, given the data available to you).

I've gone through many examples in this book where the physical and epistemic probabilities are effectively identical (especially in this chapter). This will often be true of straightforward statistical observations, especially when we are the witness in question (as then the only way the truth can deviate from what is implied by the observed physical frequencies is if there is something wrong with us or our data, e.g., we have counted wrong or are hallucinating, which we can usually rule out to a high probability—hence the importance of public or replicable data, per my first axiom in chapter 2, page 20, and my remarks earlier in this chapter, page 208) or the competent reporting witnesses are extremely numerous (e.g., the scientific community, which entails the probability of mass error or deceptive collusion is extremely small; as I remarked earlier in this chapter; I gave some examples in chapters 2 and 3, pages 28 and 42).

Mathematically: if P(r) = the probability we (or our sources) are *not* in error about the data, P(d) = the physical probability we believe is established by the data, P(d′) = what the actual physical probability is if we're wrong, and P(h|b) = the epistemic probability that we are deriving from that data, then we have to admit that P(h|b) = [P(r) × P(d)] + [(1–P(r)) × P(d′)]; so if P(r) is extremely high (i.e., if P((r)→1), then P(h|b)→P(d).[42] For example, if the true probability is (or, so far as we know, might be) 40% but we believe it's 20% and we are right about this sort of thing 99.9% of the time, then *epistemically* there is a small chance it's 40% and a large chance it's 20%, producing the conclusion that P(h|b) = (0.999 × 0.20) + (0.001 × 0.40) = 0.2002, so our estimate of 0.20 was nearly spot on. In fact, if you added in all possible probabilities (every P(d′) possible along with its probability of being the correct one), you'd probably end up with a result even nearer to 0.20.

This still only means our epistemic probability approximates what we *believe* to be the physical probability, but even so, in the given example that belief will be wrong (as in this case it is) only one in a thousand times. And that's good enough for history—especially if we are using *a fortiori* estimates. It's because of this, I suspect, that scientists sometimes confuse epistemic probabilities in BT with physical probabilities (because in their line of work, these are typically so near to identical that there is never a need to distinguish them); unlike historians, who routinely deal with fabricated or uncertain data, and thus for whom physical and epistemic probabilities often deviate considerably. See, for example, my mathematical note in chapter 3 (page 51), regarding the relationship between frequency of events and believability of testimony. Yet even there it is still a physical frequency we are talking about, namely, the frequency of certain stories being true.

Let's return to the lottery example from earlier, only this time taking it as a given that Joe is rich and, as in the Matthias case, only examining hypotheses as to how he got that way, and testing them against some body of evidence. In an ideal world (where we are omniscient and infallible), all the epistemic probabilities we assign would be exactly equal to the physical frequencies of each possibility.[43] For example, if we knew for sure that there were 100 people in MC who got rich, and we knew for sure that 2 and only 2 of them got rich by winning the CSL, then the frequency of that happening (and hence its prior probability for any person in that reference class, i.e., any person who got rich in MC) would be 1 in 50. If asked what your "degree of confidence" then was that Joe got rich by winning the lottery (prior to getting a look at any more specific evidence for that conclusion), you would have to say 2%. You couldn't say more (as you know the odds are no greater), and you couldn't say less (as you know the odds are no less). Because our information is that good, only a 2% degree of certainty is warranted by the evidence. Obviously we are almost never in such a privileged position. Our knowledge is greatly constrained, uncertain, and fallible—especially in the context of ancient history. That's why Bayesian analysis does not magically determine "the truth," all it does is demonstrate by formal logic what we are obliged to believe *given what we know at the time.* But when we pick a prior probability, we are still estimating what the physical frequency is in the most applicable reference class, and our estimates vary in reliability depending on how much relevant knowledge we have and how reliable that knowledge is.

Just as in the standard die example (which entailed a confidence of 17% that the next roll will be a '1'), if we had this perfect knowledge we could never legitimately say the prior probability of "Joe got rich by winning the lottery" was anything other than 0.02. Hence our degree of belief (prior to considering the evidence) that Joe got rich by winning the CSL would simply have to be 0.02. We could only ever say otherwise if we had to account for the fact that we lack perfect knowledge. And even then all we'd do is create an error margin around what we conclude (from what little we know) is most likely the true frequency we'd observe if we *had* perfect knowledge (and if we do this right, the odds will be high that 0.02 will fall somewhere within that confidence interval). This can be disguised but never avoided by using language about "confidence" or "certainty" or "degree of belief." If I say I am "very certain" that x, and that therefore $P(x|b) = 0.95$, I'm still really just saying that in all relevantly similar situations, the frequency of x will be "very high" relative to the frequency of ~x. I could not legitimately mean anything else, because if, in actual fact, in all relevantly similar situations the frequency of x is "very low" relative to the frequency of ~x, if I then still said I am "very certain" that x and therefore $P(x|b) = 0.95$, I would be wrong (abysmally wrong), and as soon as I had the relevant information (regarding the actual frequency of x relative to ~x), I would know I was wrong. I would then be forced to revise my certainty to reflect that better estimate of the true frequency of x. And that is always a physical frequency.

Thus, any time you talk about degrees of belief or certainty, just think about what you base that judgment on, and what facts would change your mind. Always at root you will find some sort of physical frequency that you were measuring or estimating all along. Bayesians often don't see this because they think the only alternative to subjective "degrees of certainty" is the physical frequency of some event merely happening, like the frequency of drawing a royal flush at poker, which Bayesians know can't be correct (as shown earlier, we know the prior probability of such a draw being fair is never the mere frequency of such draws—in fact, it can be very much higher than that, see page 254). But that's the wrong physical frequency. The right physical frequency is that of the most relevant cause→effect pairing, like the frequency of royal flushes being drawn fairly in relation to all royal flushes drawn. When a Bayesian says the prior probability that a royal flush is fair is 95%, because they are "very confident" that such a draw would be fair on that occasion, they are *really* saying that 95% of all royal flushes drawn (on relevantly similar occasions) are fair. Which is a physical frequency. Thus, epistemic probabilities always derive from physical probabilities. With epistemic probabilities, we're always trying to guess some actual physical probability, and thus the more we know about that physical probability, the more we will revise our epistemic probability accordingly. And as a result, epistemic probability always converges on physical probability as information approaches totality. In other words, epistemic probabilities are just physical probabilities adjusted for ignorance.

Bayesians aren't the only ones who can be confused about this.

Historians might need help understanding it, too. In our personal correspondence, C. B. McCullagh observed that to apply BT to questions in history

> the hypothetical event has to be considered as a generic type, similar in some respect to others. That might worry historians, whose hypotheses are so often quite particular. For instance, consider how the hypothesis that Henry planned to kill William II in order to seize his throne explains the fact that after his death Henry quickly seized the royal treasure. The relation between these events is rational, not a matter of frequency.[44]

But, in fact, if the connection alleged is rational, then by definition it *is* a matter of frequency, entailed by a hypothetical reference class of comparable scenarios. To say it is rational is thus identical to saying that in any set of relevantly similar circumstances, most by far will exhibit the same relation. If we didn't believe that (if we had no certainty that that relation would frequently obtain in any other relevantly similar circumstances), then the proposed inference wouldn't be rational. Explaining why confirms the point that all epistemic probabilities are approximations of physical frequencies.

The evidence in this case is that Henry not only seized the royal treasure with unusual rapidity, but that his succeeding at this would have required considerable preparations *before* William's death, and such preparations entail foreknowledge of that death. Already to say Henry seized the royal treasure "with unusual rapidity" is a plain statement of frequency, for unusual = infrequent, and this statement of frequency is either well-founded or else irrational to maintain. And if that frequency is irrational to maintain, we are not warranted in saying anything was unusual about it. Likewise, saying "it would have required considerable preparations" amounts to saying that in any hypothetical set of scenarios in all other respects identical, successful acquisition of the treasure so quickly will be infrequent, and thus improbable, *unless* prior preparations had been made (in fact, if it is claimed such success would have been *impossible* without those preparations, that amounts to saying *no* member of the reference class will contain a successful outcome *except* members that include preparations). Again, the result is said to be unusual without such preparations, or even impossible; and unusual = infrequent, while impossible = a frequency of zero. Hence such a claim to frequency must already be defensible or it must be abandoned. Similarly for every other inference: making preparations in advance of an unexpected death is inherently improbable for anyone not privy to a conspiracy to arrange that death, and being privy to such a conspiracy is improbable for anyone not actually part of that conspiracy, and in each case we have again a frequency: we are literally *saying* that in all cases of foreknowing an otherwise unpredicted death, most of those cases will involve prior knowledge of a planned murder, and in all cases of having foreknowledge of a planned murder, few will involve people not part of that plan. If *those* frequency statements are unsustainable, so are the inferences that depend on them. And so on down the line.

Thus even so particular a case as this reduces to a network of generalized frequencies. And all our judgments in this case necessarily assume we know what those frequencies are (with at least enough accuracy to warrant confidence in the conclusion). We won't know *exactly* the frequencies involved, but we know they must be generally in the ballpark stated, otherwise we wouldn't be making a rational inference at all. Hence to say that it's unlikely that this evidence would exist (the unusually rapid seizure, requiring plans entailing foreknowledge of an otherwise unexpected death) unless the proposed cause was in place (Henry conspiring to murder William), is literally identical to saying this network of frequencies must pan out as just enumerated. For if it didn't, then we would have no basis for saying any of this was unlikely. Or putting it more simply, to say it's unlikely that this evidence would exist (the surprising events that unfolded) unless the proposed cause were in place (murder) is literally to say that in any hypothetical set of scenarios in which all the same prior conditions are met but *not* that proposed cause, few will exhibit the same effects (i.e., in few of those scenarios will we see the surprising events that actually did unfold). Because this is what it *means* to say that those effects were unlikely but for the hypothesized cause; and only if those effects were unlikely but for the hypothesized cause will any confidence in that hypothesis be warranted. That's assuming its prior probability is sufficient, of course, but here we are only concerned with the consequent probability. If h = "Henry conspired in the death of William" then $\sim h$ = "Henry did not conspire in the death of

William," and then McCullagh is in effect saying (if we grant the argument he describes) that P(e|h.b) is high and P(e|~h.b) is low. And what it *means* to say the evidence makes that hypothesis likely is precisely that P(e|h.b) is high and P(e|~h.b) is low, i.e., the consequent probability of that evidence on the hypothesis "Henry did not conspire in the death of William" is low, whereas the consequent probability of that same evidence on the hypothesis "Henry did conspire in the death of William" is high.

Hence, if it's ever rational to expect a causal relation like this (with enough regularity that we can actually infer that cause from the observed effects), then that is the same thing as saying that that causal relation will frequently obtain in relevantly similar circumstances. The "relevantly similar" is defined by whatever it is (the abstracted features of the specific case) from which you infer your causal relation. As in this case, we have people profiting from deaths by acting with unusual prescience in response to those deaths. And again, this means not just actual cases (especially as often we may in fact have none), but all conceivable hypothetical cases, for example, hypothetical people of the same established means and character put in similar situations. All prediction of human behavior consists of making hypothetical estimates of the frequency of different behaviors in hypothesized conditions, and any statement about how Henry must have probably behaved before William's death given what then happened after that death is a prediction of human behavior (retroactively applied)—in this case Henry's behavior specifically, but that inference necessarily depends on assumptions about human behavior generally (generally speaking, people don't have psychic or telepathic powers; generally speaking, people don't often accidentally "discover" a murder plot they weren't involved in; etc.). And though we obviously won't have exact data from which to construct these frequencies, we know enough about what's usual and unusual (in other words, what's "frequent" and "infrequent") to develop credible *estimates* of those frequencies, especially if we argue *a fortiori* (as explained in chapter 3, page 85). Thus even here, epistemic probabilities are merely attempted estimates of physical frequencies.

The output of BT is thus what we are warranted in believing given as much as we know about all the relevant physical frequencies involved.

Accordingly, many criticisms of Bayes's Theorem are answered when BT is formulated as a theory of warrant rather than "truth." That is, BT tells us what we are warranted in believing, not what is "true" in any absolute sense. Though it also tells us we are warranted in believing that what we are warranted in believing is also true, it does so only probabilistically, that is, it guarantees that some of our beliefs will be false, but limits how many false beliefs we nevertheless have warrant to believe we have, and in that respect it's at least as adequate as any other method of forming sound beliefs, since there is no other method that guarantees that *none* of our beliefs will be false.

Treating BT as a theory of warrant, for example, resolves the problem John Earman ends with in chapter 5 of *Bayes or Bust?*: that we can never know all logically possible theories that can explain *e* and thus we can never know what their relative priors would be, yet they must all sum to 1, which would seem to leave us in a bind. But we are always and only warranted in believing what we know and what is logically entailed by what we know. Thus we don't need to know all possible theories or all their priors. Hence, just as I explained in chapter 3 (page 86), BT actually solves this problem of underdetermination: P(h|b) is by definition a probability conditional on *b*, and *b* only contains the hypotheses we know; ergo, since P(h|b) is *not* conditional on hypotheses we don't know, hypotheses we don't know have no effect on P(h|b). Until we actually discover a theory we didn't think of before: only then might that new information warrant a revision of our knowledge. But that's exactly what anyone would have expected. Hence it presents no problem. The possibility that that would happen was already mathematically accounted for in the measure of our uncertainty in assigning P(h|b), that is, our confidence level and interval for P(h|b). In other words, we already acknowledged some nonzero probability that there is some conclusion-changing hypothesis we hadn't thought of yet. Thus that we find one does not contradict our earlier assertion that there was none, because we only made that assertion in terms of probability, and since BT only tells us what we are warranted in believing with the information we have, and we didn't have that information then, its old conclusion is not contradicted by the new one, merely replaced by it (i.e., the new conclusion does not *logically* contradict the old one because the content of *b* or *e* is not the same between the two equations, hence they remain consistent). So there is no problem of logical consistency, either.

The same goes for Earman's objection that "the fact that 'old evidence' can give better confirmational value than 'new evidence' poses a major problem for Bayesianism," using the example (regarding Einstein) of the previously known "data about Mercury's perihelion" and the subsequently known "data about the bending of star light during a solar eclipse," the former being (as he argues) a stronger confirmation of Relativity Theory than the latter, yet it came *before* Einstein even formulated his theory.[45] This is not really a problem, because by itself BT is atemporal. 'When' *e* is acquired *per se* is irrelevant, as it is nowhere a factor in BT. Unless, of course, it *is* a factor—but in those cases when 'time of discovery' makes a difference, as, for example, when a lost document is 'conveniently' discovered at just the right moment in a trial, or when a prior discovery relevantly caused or influenced the subsequent discovery (though such causal influence is not *always* relevant), that in itself would become a fact in *e*, and BT would fully and correctly account for it (mathematically, this is well recognized as a dependent probability).

Otherwise, 'time of discovery' as a fact by itself simply entails identical consequent probabilities, producing a coefficient of contingency for both P(e|h.b) and P(e|~h.b) that is the same and thus cancels out without effect (as shown earlier in this chapter for all other irrelevant contingencies, see page 214). Thus the same results can often be gotten without any attention to 'when' the evidence appeared. Seen in this light, Bayes's Theorem explains the Relativity case without difficulty: we know P(e|~h.b) for the perihelion data is much lower than for the bent light data, and would remain so regardless of which data was acquired first, hence the old evidence *is* a stronger confirmation, and that it's older is simply a historical contingency of no relevance to the theory.[46]

Thus when scientists propose that "subsequent" evidence is better evidence, they are not always correct. But they often are. Because they are legitimately trying to avoid retrofitting, the tailoring of a theory to fit the evidence, and since that cannot be done in advance of unknown data, a future confirmation *can* be a stronger confirmation than a past one, because humans didn't know that that data would appear; that is, it was purely a prediction of h and neither chance nor design can credibly explain its coinciding with theory (hence entailing an extremely low consequent probability on ~h). In contrast, past evidence may have caused a theory to be tailored to explain it, thus disguising the fact that there may be better theories out there. The success of Ptolemy's geocentric model of the solar system thus disguised its error by being so well designed to "fit" previous evidence (a vast database of planetary data).[47] But Bayes's Theorem actually *better* incorporates this reasoning than any other method has done, by ensuring it only has the effect it really ought, which in some cases is exactly none. If certain data could not be predicted *except* by h, then p(e|~h) is necessarily very low (its probability approaches chance). It could not be otherwise or else it could have been predicted without h. But a set of data that precedes a theory we develop often has a greater chance of being the result of some *other* theory being true (one we didn't think of or weren't motivated to posit), and therefore p(e|~h) must be higher. This is because b includes our knowledge that while humans are not supernaturally able to predict completely unexpected phenomena (and doing so by accident is often absurdly improbable), humans are notoriously prone to overlooking or discounting alternative explanations of already-existing data, and developing 'just so' stories and other retrofitted explanations of what we observe instead (even when we try really hard not to and are "sure" we haven't). This danger of retrofitting is also controlled by careful attention to logical coherence in assigning priors, because gerrymandered theories must necessarily have low priors, especially if their components have minimal basis in actual background knowledge (see my discussion of this fact in chapters 3 and 4, pages 80 and 104). But not all false theories have to be gerrymandered to fit the evidence—particularly when the evidence is a single variable or consists of only a few variables. The more diverse and complex the observable consequences of h, the more ~h will have to be gerrymandered to produce the same consequent probability—and thus the converse follows: the less diverse and complex the observables, the less gerrymandering a false theory needs.

In the perihelion case, Relativity Theory was not created to explain that data. For b includes our knowledge of what actually *did* lead to Einstein's contrivance of the theory, which was the completely unrelated matter of conundrums in the laws of electromagnetism and the velocity of light, and once Einstein developed Special Relativity to account for *that*, he then expanded Special Relativity into General Relativity in order to account for the physics of acceleration, and only then tested *that* on the perihelion data. Which means he did not gerrymander Relativity Theory to explain the perihelion data, but quite the contrary: he could not possibly have predicted that a solution to the problem of the constant velocity of light would perfectly explain oddities in the orbit of Mercury. That it could do this did become apparent in the course of his development of Special Relativity into General Relativity, but by then the latter was already logically entailed by the former (in conjunction with the known physics of acceleration). So in fact, that Relativity Theory would explain the perihelion data was just as unexpected after the perihelion data was acquired as it would have been had that data only been acquired after the theory was formulated.

So the danger of retrofitting was no real concern in that case. Understood in terms of what we are warranted in believing, scientists were warranted in believing Relativity was validly confirmed by that previously collected data for the orbit of Mercury (certainly given the fact that that test alone would never have been regarded as sufficient anyway, and never was). Or to put it another way, in that particular case, the probability that retrofitting caused the correlation (rather than the theory's being true) is extremely small. But since that is not the case for many other scientific hypotheses (and, indeed, most hypotheses about history), often future evidence does carry more weight than past evidence. But that fact will be fully represented in BT. Hence note that this does not ever mean past evidence carries *no* weight, even in cases with the greatest risk of retrofitting. Evidence cannot be ignored. *Everything* we know must be included in either *e* or *b*, thus *h* must either account for all past evidence, or else past evidence must affect the probability of new evidence on *h* and ~*h* (i.e., past evidence must at the very least become part of our background knowledge).[48] In other words, the "problem of old evidence" cannot be used as an excuse to ignore old evidence.

To tie this into all the preceding, this consideration of the probability that retrofitting explains a successful test rather than a theory's being true, is again just another frequency: the frequency with which explanations bearing a certain relationship to past evidence will be retrofitted rather than true. When that "certain relationship" is like that of the perihelion case, this frequency will be extremely small (in all comparable cases, retrofitting is almost never going to be the explanation), but when it's like that of Ptolemy's geocentrism, this frequency will be high enough to require taking it into account (as in all comparable cases, retrofitting will often be the explanation, or at least often enough to take note of). In such cases 'when' data is acquired becomes relevant because there is a possible causal relationship at play, causal relationships are temporal, and e and b contain all known causal relationships. In Jesus studies, this is reflected in the frequency with which theories affirmed about Jesus just happen to track the theological and cultural and even personal or political interests of the historians affirming them (see note 23 for chapter 1, page 296). This entails the frequency of retrofitting there is high. But well-constructed hypotheses possessing optimum simplicity and well supported by existing evidence according to a sound BT analysis should minimize the risk of retrofitting, thereby reducing "the frequency with which explanations like ours will be retrofitted rather than true" to such a degree that the possibility of retrofitting will be washed out by our *a fortiori* estimates and can therefore be ignored.

CONCLUSION

This chapter has been rather technical, but its aim was to address the most sophisticated critical concerns about BT that relate not only to the validity of its application to history but the actual mechanics of applying it to history. Some critics object that using BT makes it harder to resolve disagreements because disagreements over probabilities are subjective, but arguing from my first section above (in conjunction with my analysis in chapter 3, page 81), we can conclude that BT does not make anything harder but in fact more transparent. If it's hard to resolve disagreements with BT, this is because it's hard to resolve disagreements period. BT does not in fact make this more difficult; it actually makes it easier, by forcing us to own up to our unstated assumptions, thereby exposing them to criticism and forcing us to justify them in evidence (or else abandon them for assumptions that can be thus justified). In other words, BT allows opposing sides to isolate just what it really is that sustains their disagreement, thus making progress (or at least mutual understanding) possible (see pages 208–14).

Some critics object that deriving conclusions about what happened in the past by using probabilistic reasoning about how surviving evidence came about is impossible, because the specific probabilities that any actual item of evidence would survive (as opposed to none, or some other item of evidence instead) are astronomically small, and likewise even the probabilities that the evidence would come to exist in the first place. I met this objection by demonstrating that all such contingencies cancel out and thus make no difference, allowing us to focus on general kinds and properties of evidence, and predictions as to type of evidence, all conditional on known processes and their effects on evidence survival (see pages 214–28). Thus the acceptance in chapter 3 (page 77) of generic rather than exact prediction was here confirmed to be valid.

Some critics object that so many problems arise in the task of assigning prior probabilities that the endeavor should seem hopeless. I addressed all the most common problems that so arise and how to resolve them, including some basic advice on how historians can go about developing defensible prior probabilities (and briefly extending the same points to developing sound consequent probabilities as well). The difficulties that remain here are no greater than are already faced by any other method, so they can present no objection to BT. Historians already rely on these assumptions about prior probability in everything they argue. Understanding how these assumptions derive from reference classes makes it possible to evaluate those assumptions (see pages 229–56).

Some critics also object to the use of hypothetical frequencies rather than sticking to hard data. I dismissed that objection as unfounded (see pages 257–65). Reasoning about probabilities in the real world often demands resorting to hypothetical frequencies, especially in reasoning about the past, which is rife with unique conjunctions of events and obscured data. That valid reasoning can nevertheless proceed has enormous relevance to historians, who depend on hypothetical frequencies even more than scientists do (whether they use BT or not). To that end I gave some advice on how to develop and use them.

Finally, some critics object to Bayes's Theorem because it entails abandoning a frequentist interpretation of probability. Since I proved it does not, but in fact *entails* a frequentist interpretation of probability, this objection is a nonstarter (see pages 265–80). Here it may be the Bayesians who will be the more appalled, but I believe I have made a persuasive case for the conclusion that *all* Bayesians are really talking about frequencies. They just sometimes don't realize what their probabilities are frequencies of. And by identifying what they really are, I demonstrated how all epistemic probabilities derive from physical frequencies (which are usually hypothetical frequencies, of course, but still frequencies). This led me to conclude with a discussion of the difference between accepting BT as a theory of warranted belief, and mistaking it as a method of arriving at indisputable truths.

Perhaps other objections will be imagined. But given the arguments of this and previous chapters, I doubt any will carry force. The conclusion seems inescapable to me. Historians should be Bayesians. Such is the general point I have proved throughout this book. Chapter 2 laid down general rules for developing conclusions about history, chapter 3 explained Bayes's Theorem and the basics of its application, and chapter 4 applied that theorem to the most common and fundamental historical methods, proving they all reduce to BT and are better informed when understood as such. But then I went beyond the general point, and in chapter 5 I applied this conclusion to the study of the historical Jesus in particular. There I showed that none of the methods now used by Jesus scholars to reconstruct the historical Jesus are logically valid or validly employed, except insofar as they can be reframed in Bayesian terms, in a logically valid and factually sound way. The special conclusion thus follows from the general: historians of Jesus should be Bayesians.

Therefore, in volume 2, *On the Historicity of Jesus Christ*, I will apply Bayes's Theorem to the total body of evidence pertaining to the historical Jesus in order to see what we can thus conclude.

APPENDIX

Here included are various specific tools in one convenient place: all the formulas for applying Bayes's Theorem; the Canon of Probabilities for easing selection of *a fortiori* probability estimates; and a flow chart for applying Bayesian reasoning without any numbers or math at all (useful only in simple cases).

FORMULAE

The short form of Bayes's Theorem reads:

$$P(h|e.b) = \frac{P(h|b) \times P(e|h.b)}{P(e|b)}$$

Which in more advanced mathematical notation reads:

$$P(h_0|e.b) = \frac{P(h_0|b)\, P(e|h_0.b)}{\sum_n P(h_n|b)\, P(e|h_n.b)}$$

But this actually represents (and mathematically translates to) the long form:

$$P(h|e.b) = \frac{P(h|b) \times P(e|h.b)}{[\ P(h|b) \times P(e|h.b)\ ]\ +\ [\ P(\sim h|b) \times P(e|\sim h.b)\ ]}$$

Which represents the logic:

"given all we know so far, then . . ."

And when comparing more than two hypotheses, the long form can be expanded to:

$$P(h_1|e.b)\ =\ \frac{P(h_1|b) \times P(e|h_1.b)}{[\ P(h_1|b) \times P(e|h_1.b)\ ]+[\ P(h_2|b) \times P(e|h_2.b)\ ]+[\ P(h_3|b) \times P(e|h_3.b)\ ]}$$

Which compares three hypotheses. If you want to compare more than three hypotheses, just add as many boxes to the denominator as you need (i.e., repeat the "$[P(h_3|b) \times P(e|h_3.b)]$" box for h_4, and so on); and remember that all $P(h_n|b)$ must sum to 1.

THE ODDS FORM

You can also represent Bayes's Theorem in the form of calculating the *odds* of something being true rather than the *probability*, and some prefer this form, as it more clearly separates the prior odds from the weight of the evidence, and shows more clearly how it is the *ratio* of probabilities that matters rather than the actual probabilities themselves; while others find this approach more complicated and confusing. But if you want to work with odds, the equation becomes:

$$\frac{P(h|e.b)}{P(\sim h|e.b)} = \frac{P(h|b)}{P(\sim h|b)} \times \frac{P(e|h.b)}{P(e|\sim h.b)}$$

To illustrate how this form would be used, if the priors are 0.75 and 0.25 in favor of h, then the "prior odds" of h are 3:1, and if the evidence is 100% expected on h but the same evidence would only result half the time on $\sim h$, then the odds favoring h are 2:1 (representing the weight of the evidence, i.e., how strongly it supports h), and then the total odds that h is true would be 6:1 (3 × 2). In other words: (0.75/0.25) × (1/0.5) = 3/1 × 2/1 = 3 × 2 = 6, which is 6/1, or just 6:1, which is P(h|e.b) in ratio to P(~h|e.b). The odds are then six to one that h is true. This form is convenient if you can more readily estimate the odds than the probabilities (e.g., if you are sure going in that h is three times more likely than $\sim h$, but you aren't sure what either prior probability actually is; or if you don't know how likely the evidence is but are sure the evidence is twice as likely on h than on $\sim h$; or both).

For convenience, the most common and useful mathematical symbols are:

= (equals)
> (greater than)
< (less than)
>> (much greater than)
<< (much less than)
>>> (very much greater than)
<<< (very much less than)
≥ (greater than or equal to)
≤ (less than or equal to)
≈ (approximately equal to)
→ (approaching)[†]

† That last symbol can also mean other things in different contexts: in contexts pertaining to causal sequence it means 'and then' or 'causes'; and pertaining to logical entailment, 'entails.'

CANON OF PROBABILITIES

"Virtually Impossible" = 0.0001% (*1 in 1,000,000*) = 0.000001
"Extremely Improbable" = 1% (*1 in 100*) = 0.01
"Very Improbable" = 5% (*1 in 20*) = 0.05
"Improbable" = 20% (*1 in 5*) = 0.20
"Slightly Improbable" = 40% (*2 in 5*) = 0.40
"Even Odds" = 50% (*50/50*) = 0.50
"Slightly Probable" = 60% (*2 in 5 chance of being otherwise*) = 0.60
"Probable" = 80% (*only 1 in 5 chance of being otherwise*) = 0.80
"Very Probable" = 95% (*only 1 in 20 chance of being otherwise*) = 0.95
"Extremely Probable" = 99% (*only 1 in 100 chance of being otherwise*) = 0.99
"Virtually Certain" = 99.9999% (*1 in 1,000,000 otherwise*) = 0.999999

BAYESIAN REASONING WITHOUT MATH

Reproduced from *Sources of the Jesus Tradition*, ed. R. Joseph Hoffmann (Amherst, NY: Prometheus Books, 2010), with a colloquial version following (on page 288):

Bayesian Flowchart

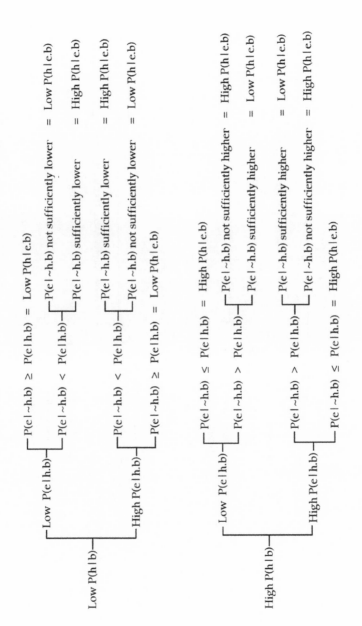

Low P(h | b)
- Low P(e | h.b)
  - P(e | ~h.b) ≥ P(e | h.b) = Low P(h | e.b)
  - P(e | ~h.b) < P(e | h.b)
    - P(e | ~h.b) not sufficiently lower = Low P(h | e.b)
    - P(e | ~h.b) sufficiently lower = High P(h | e.b)
- High P(e | h.b)
  - P(e | ~h.b) < P(e | h.b)
    - P(e | ~h.b) sufficiently lower = High P(h | e.b)
    - P(e | ~h.b) not sufficiently lower = Low P(h | e.b)
  - P(e | ~h.b) ≥ P(e | h.b) = Low P(h | e.b)

High P(h | b)
- Low P(e | h.b)
  - P(e | ~h.b) ≤ P(e | h.b) = High P(h | e.b)
  - P(e | ~h.b) > P(e | h.b)
    - P(e | ~h.b) not sufficiently higher = High P(h | e.b)
    - P(e | ~h.b) sufficiently higher = Low P(h | e.b)
- High P(e | h.b)
  - P(e | ~h.b) > P(e | h.b)
    - P(e | ~h.b) sufficiently higher = Low P(h | e.b)
    - P(e | ~h.b) not sufficiently higher = High P(h | e.b)
  - P(e | ~h.b) ≤ P(e | h.b) = High P(h | e.b)

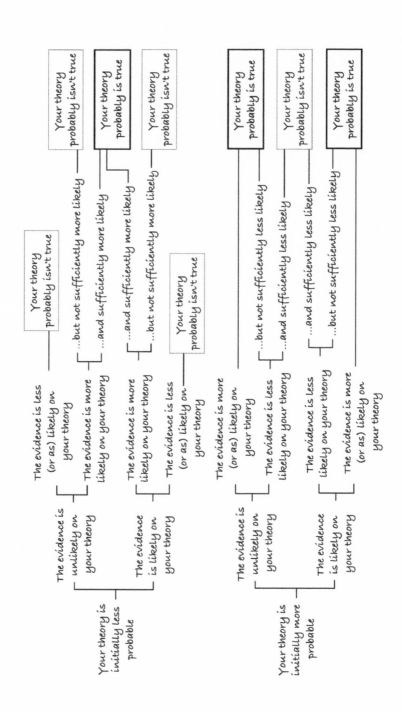

This chart (presented here in two forms, formal and informal) represents the complete logic of Bayes's Theorem (for two competing hypotheses), which can be used with nonnumerical declarations of relative likelihood at each step. All you must decide at each stage is which is more likely than the other. The answer leads you to the next decision, until you get an answer. For example, the first decision: is the hypothesis inherently more likely than not? If yes, it has a high prior; if not, a low prior. The second decision: is the evidence more likely on your hypothesis, or not? And so on.

To use either chart, the terms "Low" and "Unlikely" mean lower than 50% (< 0.50), and "High" or "Likely" mean higher than 50% (> 0.50), although when $P(e|h.b) = 50\%$ (i.e., when the evidence is no more likely than unlikely), then treat it as "High" (i.e., likely) if $P(h|b)$ is "High," and "Low" (i.e., unlikely) if "Low," accordingly. "Sufficiently lower" and "sufficiently less likely" (or "sufficiently higher" and "sufficiently more likely") mean $P(e|{\sim}h.b)$ is lower (or higher) than $P(e|h.b)$ by enough to overcome the prior probability (and thus produce a conclusion contrary to what the prior probability alone would predict), though there is hardly any principled way to determine this without returning to the math. To read the results on the formalized chart, a high $P(h|e.b)$ means your hypothesis is more likely true; a low $P(h|e.b)$, more likely false (and the higher or lower, the more likely either way). Both charts are identical, it's just that one uses the formal terms and symbols, the other, plain English.

The one circumstance not represented on the flowchart is when $P(h|b) = 50\%$ (and therefore is neither "High" nor "Low," i.e., neither hypothesis is "initially more probable" than the other). In that event the hypothesis with the higher $P(e|b)$ is more likely true (i.e., the one for which the evidence is more likely), and if both hypotheses are equal on that measure as well, then the hypothesis is as likely as not (the outcome is simply 50/50).

COMMON ABBREVIATIONS

NIF = Richard Carrier, *Not the Impossible Faith: Why Christianity Didn't Need a Miracle to Succeed* (Raleigh, NC: Lulu.com, 2009).

SGG = Richard Carrier, *Sense and Goodness without God: A Defense of Metaphysical Naturalism* (Bloomington, IN: AuthorHouse, 2005).

TCD = John Loftus, ed., *The Christian Delusion: Why Faith Fails* (Amherst, NY: Prometheus Books, 2010).

TCD$_s$ = Richard Carrier, "Christianity Was Not Responsible for Modern Science," *TCD*, pp. 396–419.

TCD$_w$ = Richard Carrier, "Why the Resurrection Is Unbelievable," *TCD*, pp. 291–315.

TEC = John Loftus, ed., *The End of Christianity* (Amherst, NY: Prometheus Books, 2011).

TEC$_d$ = Richard Carrier, "Neither Life nor the Universe Appear Intelligently Designed," *TEC*, pp. 279–304, 404–14.

TEC$_m$ = Richard Carrier, "Moral Facts Naturally Exist (and Science Could Find Them)," *TEC*, pp. 333–64, 420–29.

TEC$_s$ = Richard Carrier, "Christianity's Success Was Not Incredible," *TEC*, pp. 53–74, 372–75.

TET = Robert Price and Jeffery Lowder, ed., *The Empty Tomb: Jesus Beyond the Grave* (Amherst, NY: Prometheus Books, 2005).

TET$_b$ = Richard Carrier, "The Burial of Jesus in Light of Jewish Law," *TET*, pp. 369–92.

TET$_s$ = Richard Carrier, "The Spiritual Body of Christ and the Legend of the Empty Tomb," *TET*, pp. 105–232.

TET$_t$ = Richard Carrier, "The Plausibility of Theft," *TET*, pp. 349–68.

NOTES

CHAPTER 1. THE PROBLEM

1. For surveys of the field demonstrating this, see Bart Ehrman, *Jesus, Interrupted: Revealing the Hidden Contradictions in the Bible (and Why We Don't Know about Them)* (New York: HarperOne, 2009); Burton Mack, *The Christian Myth: Origins, Logic, and Legacy* (New York: Continuum, 2001); Gerd Theissen and Annette Merz, *The Historical Jesus: A Comprehensive Guide*, trans. John Bowden (Minneapolis: Fortress Press, 1996).

2. Quoted in Stanley Porter, *The Criteria for Authenticity in Historical-Jesus Research: Previous Discussion and New Proposals* (Sheffield, UK: Sheffield Academic Press, 2000), p. 115.

3. Ibid., pp. 116–17. See also Stanley Porter's summary critique in James Charlesworth and Petr Pokorný, eds., *Jesus Research: An International Perspective* (Grand Rapids, MI: William B. Eerdmans, 2009), pp. 16–35.

4. Dale Allison, "The Historians' Jesus and the Church," in *Seeking the Identity of Jesus: A Pilgrimage*, eds. Beverly Roberts Gaventa and Richard B. Hays (Grand Rapids, MI: William B. Eerdmans, 2008), pp. 79–95 (quoting p. 79). His conclusion has only become stronger after a decade of critical research: compare Dale Allison, *Jesus of Nazareth: Millenarian Prophet* (Minneapolis: Fortress Press, 1998), pp. 1–77.

5. Hector Avalos, *The End of Biblical Studies* (Amherst, NY: Prometheus Books, 2007), pp. 203–209; Michael Bird, "The Criterion of Greek Language and Context: A Response to Stanley E. Porter," *Journal for the Study of the Historical Jesus* 4, no. 1 (2006): pp. 55–67.

6. Stanley Porter, "The Criterion of Greek Language and Its Context: A Further Response," *Journal for the Study of the Historical Jesus* 4, no. 1 (2006): 69–74 (in response to Bird, cited in the previous note). Porter concludes his new criteria only establish the *possibility* of historicity, but that's of no use if you want to know what actually is historical. And there's more wrong with his new criteria than even Porter concedes (as I'll show in chapter 5).

7. See Dale Allison, "The Historians' Jesus and the Church" and *The Historical Christ and the Theological Jesus* (Grand Rapids, MI: William B. Eerdmans, 2009); Gerd Theissen and Dagmar Winter, *The Quest for the Plausible Jesus: The Question of Criteria*, trans.

M. Eugene Boring (Louisville, KY: John Knox Press, 2002); Chris Keith and Anthony Le Donne, eds., *Jesus, Criteria, and the Demise of Authenticity* (T & T Clark, 2002); and Porter, *Criteria for Authenticity*. Similar doubts can be found almost anywhere the criteria have ever been critically discussed, e.g., M. D. Hooker, "Christology and Methodology," *New Testament Studies* 17 (1970): pp. 480–87; John Gager, "The Gospels and Jesus: Some Doubts about Method," *Journal of Religion* 54, no. 3 (July 1974): 244–72; Eugene Boring, "The Beatitudes in Q and Thomas as a Test Case," *Semeia* 44 (1988): 9–44; John Meier, "Criteria: How Do We Decide What Comes from Jesus?" *A Marginal Jew: Rethinking the Historical Jesus*, vol. 1 (New York: Doubleday, 1991), pp. 167–95; Christopher Tuckett, "Sources and Methods," in *The Cambridge Companion to Jesus*, ed. Markus Bockmuehl (Cambridge: Cambridge University Press, 2001), pp. 121–37; H. W. Shin, *Textual Criticism and the Synoptic Problem in Historical Jesus Research: The Search for Valid Criteria* (Dudley, MA: Peeters, 2004), pp. 135–220, pp. 320–34; Eric Eve, "Meier, Miracle, and Multiple Attestation," *Journal for the Study of the Historical Jesus* 3, no. 1 (2005): 23–45; William John Lyons, "The Hermeneutics of Fictional Black and Factual Red: The Markan Simon of Cyrene and the Quest for the Historical Jesus," *Journal for the Study of the Historical Jesus* 4, no. 2 (June 2006): 139–54 (cf. 150–51, n. 51) and "A Prophet Is Rejected in His Home Town (Mark 6.4 and Parallels): A Study in the Methodological (In)Consistency of the Jesus Seminar," *Journal for the Study of the Historical Jesus* 6, no. 1 (March 2008): 59–84; and Rafael Rodríguez, "Authenticating Criteria: The Use and Misuse of a Critical Method," *Journal for the Study of the Historical Jesus* 7, no. 2 (2009): 152–67. The discussion of the same criteria in the Jesus Seminar's manual on method, edited by Bernard Brandon Scott, *Finding the Historical Jesus: Rules of Evidence* (Santa Rosa, CA: Polebridge, 2008), is almost wholly uncritical and entirely unresponsive to any of the literature above.

8. Although see Fernando Bermejo-Rubio, "The Fiction of the 'Three Quests': An Argument for Dismantling a Dubious Historiographical Paradigm," *Journal for the Study of the Historical Jesus* 7.3 (2009): 211–53, who calls into question the entire paradigm of numbering and distinguishing three separate "quests" for the historical Jesus in this excellent study of the history of the three quests that will be extremely informative even for those who disagree with its thesis—in fact, I believe it should become required reading in the field.

9. For various demonstrations of this failure, see R. Joseph Hoffmann, ed., *Sources of the Jesus Tradition: Separating History from Myth* (Amherst, NY: Prometheus Books, 2010) and Thomas Verenna and Thomas L. Thompson, eds., *'Is This Not the Carpenter?' The Question of the Historicity of the Figure of Jesus* (Sheffield, UK: Equinox, 2012). But the works in the following note already document and exemplify the problem.

10. For some of the best (although one must read them all in light of Bermejo-Rubio, "The Fiction of the 'Three Quests,'" cited in n. 8) see Dale Allison, *The Historical Christ*; David Gowler, *What Are They Saying about the Historical Jesus?* (New York: Paulist,

2007); Mark Strauss, *Four Portraits, One Jesus: An Introduction to Jesus and the Gospels* (Grand Rapids, MI: Zondervan, 2007), pp. 358–82 (with pp. 397–98 and p. 491); Ben Witherington III, *What Have They Done with Jesus? Beyond Strange Theories and Bad History* (New York: HarperOne, 2006), sounding a more desperate note than in his earlier but still relevant account of Jesus studies' failures in *The Jesus Quest: The Third Search for the Jew of Nazareth* (Downers Grove, IL: InterVarsity, 1997); Robert Price, *The Pre-Nicene New Testament: Fifty-Four Formative Texts* (Salt Lake City: Signature Books, 2006), pp. 1169–80; Craig Evans, *Fabricating Jesus: How Modern Scholars Distort the Gospels* (Downers Grove, IL: InterVarsity: 2006); James Dunn and Scot McKnight, eds., *The Historical Jesus in Recent Research* (Winona Lake, IN: Eisenbrauns, 2005); Craig Evans, ed., *The Historical Jesus: Critical Concepts in Religious Studies*, vol. 1: *The History of the Quest: Classical Studies and Critical Questions* (London: Routledge, 2004); Donald L. Denton, *Historiography and Hermeneutics in Jesus Studies: An Examination of the Work of John Dominic Crossan and Ben F. Meyer* (New York: T & T Clark, 2004); Darrell Bock, *Studying the Historical Jesus: A Guide to Sources and Methods* (Grand Rapids, MI: Baker Academic, 2002); Markus Bockmuehl, *Cambridge Companion to Jesus*, pp. 121–83; Bruce Chilton and Craig Evans, eds., *Studying the Historical Jesus: Evaluations of the State of Current Research* (New York: E. J. Brill, 1998); Mark Allen Powell, *Jesus as a Figure in History: How Modern Historians View the Man from Galilee* (Louisville, KY: John Knox, 1998); Raymond Brown, *An Introduction to the New Testament* (New York: Paulist 1997), pp. 817–30; Luke Timothy Johnson, *The Real Jesus: The Misguided Quest for the Historical Jesus and the Truth of the Traditional Gospels* (New York: HarperSanFrancisco, 1996); Marcus Borg, *Jesus in Contemporary Scholarship* (Valley Forge, PA: Trinity, 1994); with additional surveys in many of the previously cited references on historicity criteria, for example, Porter, *Criteria for Authenticity*, pp. 28–62 (with even more discussion and bibliographies throughout pp. 63–125). And this is not even remotely a complete list. For more detail on the history of the quest in earlier eras (which has obvious similarities to more recent developments), see Walter P. Weaver, *The Historical Jesus in the Twentieth Century, 1900–1950* (Valley Forge, PA: Trinity, 1999) and Albert Schweitzer, *The Quest of the Historical Jesus: A Critical Study of Its Progress from Reimarus to Wrede*, trans. W. Montgomery (New York: Macmillan, 1910).

11. Strauss, *Four Portraits*, pp. 365–77.

12. Ibid., pp. 366–68.

13. Ibid., pp. 368–69.

14. Ibid., pp. 369–71.

15. Ibid., pp. 372–77.

16. Ibid., p. 377.

17. Alvar Ellegård, *Jesus—One Hundred Years before Christ: A Study in Creative Mythology* (Woodstock, NY: Overlook, 1999); Israel Knohl, *The Messiah before Jesus:*

The Suffering Servant of the Dead Sea Scrolls (Berkeley: University of California Press, 2000); and Robert Eisenman, in a whole career of books: *The Dead Sea Scrolls and the First Christians* (Edison, NJ: Castle Books, 2009), *The New Testament Code: The Cup of the Lord, the Damascus Covenant, and the Blood of Christ* (London: Watkins, 2006), and *James the Brother of Jesus: The Key to Unlocking the Secrets of Early Christianity and the Dead Sea Scrolls* (New York: Viking, 1997).

18. Scot McKnight, *Jesus and His Death: Historiography, the Historical Jesus, and Atonement Theory* (Waco, TX: Baylor University Press, 2005).

19. William Herzog II, *Prophet and Teacher: An Introduction to the Historical Jesus* (Louisville, KY: John Knox, 2005), p. 12.

20. Thomas Verenna, *Of Men and Muses: Essays on History, Literature, and Religion* (Raleigh, NC: Lulu.com, 2009), pp. 46–47, gives an even longer list, with references; likewise Verenna and Thompson, "Introduction," *'Is This Not the Carpenter?'* (cf. pp. 9–10). Another list (and the same conclusion) is provided in John Dominic Crossan, *The Historical Jesus: The Life of a Mediterranean Jewish Peasant* (New York: HarperSanFrancisco, 1992), p. xxviii. For a detailed study of this scandalous diversity of views, see Powell, *Jesus as a Figure in History.*

21. Helmut Koester, "The Historical Jesus and the Historical Situation of the Quest: An Epilogue," in Chilton and Evans, *Studying the Historical Jesus,* pp. 535–45 (quote from p. 544).

22. Charlesworth and Pokorný, eds., *Jesus Research: An International Perspective,* p. 1.

23. For example, see John C. Poirier, "Seeing What Is There in Spite of Ourselves: George Tyrrell, John Dominic Crossan, and Robert Frost on Faces in Deep Wells," *Journal for the Study of the Historical Jesus* 4, no. 2 (2006): 127–38, and James Crossley, *Jesus in an Age of Terror: Scholarly Projects for a New American Century* (Oakville, CT: Equinox, 2008). Of course *popular* views of the historical Jesus deviate much farther from even the many Jesuses reconstructed by expert historians, and these also track what people want Jesus to have been rather than what he really was, but that won't be my concern here. For example, see Stephen Prothero, *American Jesus: How the Son of God Became a National Icon* (New York: Farrar, Straus and Giroux, 2003) and Adele Reinhartz, *Jesus of Hollywood* (Oxford: Oxford University Press, 2007). Here my concern is with what should be well-founded expert historical conclusions about Jesus, not popular, dogmatic, or religious conclusions about Jesus.

CHAPTER 2. THE BASICS

1. For the epistemology underlying these axioms and the concepts and assumptions within them, see my book *SGG* (esp. pp. 21–62 and pp. 211–52) and my essay

"Epistemological End Game" (November 29, 2006) at http://richardcarrier.blogspot.com/2006/11/epistemological-end-game.html. I further discuss the epistemology of history in my essay "Experimental History" (June 28, 2007) at http://richardcarrier.blogspot.com/2007/07/experimental-history.html. These issues are also discussed to some degree by other scholars of historical method, as listed in note 3 for chapter 4, page 306.

2. It's therefore scandalous that historians typically do not even study logic. The perils of this neglect are thoroughly documented by David Hackett Fischer, *Historians' Fallacies: Toward a Logic of Historical Thought* (New York: Harper & Row, 1970).

3. See Richard Carrier, "The Function of the Historian in Society," *History Teacher* 35, no. 4 (August 2002): 519–26.

4. Note that I am deliberately subverting the usual convention of calling these "objective" and "subjective" probabilities, respectively. That distinction is actually confusing and illogical and should be avoided. Here I shall only use those terms to denote what they usually mean in epistemological theory: subjective probabilities are estimates based on how things seem to the individual estimator, and objective probabilities are the actual probabilities individuals are trying to estimate. This convention is to be preferred because all probabilities, even "subjective" or "epistemic" probabilities, reduce to physical frequencies, as I'll demonstrate in chapter 6 (pp. 265–80). Hence, objective probabilities are "true" or "false," but subjective probabilities are "close to" or "far from" some true probability.

5. If P(m) is the probability of any single expert missing an error and P(1000m) is the probability of a thousand such experts missing that same error, and they are all acting independently, then when P(m) = x, then P(1000m) = $x^{1,000}$, which is extraordinarily small, but still not zero. For example, even if x were as extraordinarily high as 0.5 (i.e., an error is missed fully half the time), $x^{1,000}$ would still be less than 1 in 10^{301}, the latter a number of ghastly size (a one followed by over three hundred zeroes), entailing an unimaginably small fraction. Yet that's still not zero. Note that this does not mean if everyone agrees, then they must be right, because we must also account for causes of their agreement other than the examined claim being true (and how to do that I will explain in chapter 3). For example, in Condorcet's Jury Theorem (which holds that if the probability of an expert being wrong is less than 0.5, then the majority opinion of a pool of experts will be correct to a probability exceeding 0.5 and approaching 1 as the number of experts increases), the base probability of an "expert" being "correct" only really reflects the degree to which an expert opinion is "correctly caused" (and according to Condorcet's theorem, when that probability falls below 0.5, a majority consensus becomes increasingly *incorrect*) and furthermore assumes experts all decide independently of each other (which is rarely if ever true). Thus to apply Condorcet's theorem we would have to rule out errors shared by all experts (such as a shared bias, or a shared method that's logically invalid), and this is what makes critical surveys of expert biases and methods so crucial.

6. Although I believe the traditional definition can be preserved if all knowledge claims

are stated as probabilities: if all propositions of the form K ("I know that x") translate to Kp ("I know that probably x"), and Kp derives from data sampling all the way down to raw uninterpreted experience (which as such has a zero probability of being false), then Kp (and thus also K) can be "justified true belief" even when x is false. For this reason I believe the only valid or useful epistemology is probabilistic, wherein all claims K translate to claims Kp (except claims of raw uninterpreted experience). But demonstrating that is beyond the scope of the present work.

7. For discussion of this point using a real example, see TET_s, pp. 180–82.

8. Of course, epistemologically, the difference between facts and theories is, like the difference between mountains and hills, a matter of degree: apart from the undeniables of immediate experience, all facts are theoretical. Even if you are dreaming or hallucinating the page now before you, it is still a fact that you are now seeing a page before you, but it is a "theory" that you are reading a real physical book and not dreaming or hallucinating it. It's just that this is a theory we conclude has an extremely high probability of being true, so high that we designate it a "fact," in order to distinguish it from theories we must be less certain of.

9. I make the most detailed case for this in TET_s, especially on pp. 155–97. Many scholars agree with me in significant detail, including James Tabor, "Leaving the Bones Behind: A Resurrected Jesus Tradition with an Intact Tomb," paper presented in 2008 at the "Sources of the Jesus Tradition" conference in Amherst, NY; Bruce Chilton, *Rabbi Paul: An Intellectual Biography* (New York: Doubleday, 2004), pp. 57–58; Gregory Riley, *Resurrection Reconsidered: Thomas and John in Controversy* (Minneapolis: Fortress 1995); and Adela Collins, "The Empty Tomb in the Gospel According to Mark," *Hermes and Athena: Biblical Exegesis and Philosophical Theology*, ed. Eleonore Stump and Thomas Flint (Notre Dame, IN: University of Notre Dame Press, 1993), pp. 107–40. Support is also found in Peter Lampe, "Paul's Concept of a Spiritual Body," *Resurrection: Theological and Scientific Assessments*, Ted Peters, Robert John Russell, and Michael Welker, ed. (Grand Rapids, MI: William B. Eerdmans, 2002), pp. 103–14; Dale Martin, *The Corinthian Body* (New Haven, CT: Yale University Press, 1995); and C. F. Moule, "St. Paul and Dualism: The Pauline Conception of the Resurrection," *New Testament Studies* 12 (1966): 106–23. Many more historians (far too many to name) likewise agree that the empty tomb story is a legend or could be a legend.

10. This was an actual debate, which has been conclusively resolved from extensive examination of the evidence (comprising a good example of the complexity of generalizations and the evidence required to establish them): everyone in antiquity read silently when they wanted to, read aloud as performance and entertainment more frequently than we do, and just as we read aloud sometimes even to ourselves, so did they. See William Johnson, "Toward a Sociology of Reading in Classical Antiquity," *American Journal of Philology* 121, no. 4 (Winter 2000): 593–627; A. K. Gavrilov, "Reading Techniques in

Classical Antiquity," *Classical Quarterly* 47 (1997): 56–73; M. F. Burnyeat, "Postscript on Silent Reading," *Classical Quarterly* 47 (1997): 74–76; and Bernard Knox, "Silent Reading in Antiquity," *Greek, Roman, and Byzantine Studies* 9 (1968): 421–35.

11. I explain why in Richard Carrier, "History Before 1950," April 30, 2007, at http://richardcarrier.blogspot.com/2007/04/history-before-1950.html.

CHAPTER 3. INTRODUCING BAYES'S THEOREM

1. John Earman, *Bayes or Bust? A Critical Examination of Bayesian Confirmation Theory* (Cambridge, MA: MIT Press, 1992), p. 117.

2. On the evidence for Thallus and its problems, see Robert Van Voorst, *Jesus Outside the New Testament: An Introduction to the Ancient Evidence* (Grand Rapids, MI: William B. Eerdmans, 2000), pp. 20–23. But many of his inferences are fallacious or naive: contrast his conclusions with the more astute analysis of Felix Jacoby, *Fragmente der griechischen Historiker* (Leiden: Brill, 1954), § 256 F1 (for a complete translation and commentary thereof, see Richard Carrier, "Jacoby and Müller on 'Thallus,'" Secular Web, 1999, at http://www.infidels.org/library/modern/richard_carrier/jacoby.html); and see Richard Carrier, "Thallus and the Darkness at Christ's Death," *Journal of Greco-Roman Christianity and Judaism* (forthcoming).

3. For an extensive list, see n. 3 for chap. 4 (page 306).

4. For examples and discussion of this kind of 'scientific history,' see Matthew Hedman, *The Age of Everything: How Science Explores the Past* (Chicago: University of Chicago Press, 2007). See also Cynthia Stokes Brown, *Big History: From the Big Bang to the Present* (New York: New Press, 2007) and Eric Chaisson, *Epic of Evolution: Seven Ages of the Cosmos* (New York: Columbia University Press, 2006).

5. For demonstrations of why the preceding analysis is philosophically sound, see Massimo Pigliucci, *Nonsense on Stilts: How to Tell Science from Bunk* (Chicago: University of Chicago Press, 2010), pp. 18–23 and pp. 45–55; Marc Trachtenberg, *The Craft of International History: A Guide to Method* (Princeton, NJ: Princeton University Press, 2006), pp. 14–29; and (most extensively and valuably) John Lewis Gaddis, *The Landscape of History: How Historians Map the Past* (New York: Oxford University Press, 2002). I will say more about the equivalence of history and science in chapter 4, where I discuss the hypothetico-deductive method (p. 104).

6. The present book greatly expands on and perfects my earlier arguments in support of the same conclusion: Richard Carrier, "Bayes's Theorem for Beginners: Formal Logic and Its Relevance to Historical Method," *Sources of the Jesus Tradition: Separating History from Myth*, ed. R. Joseph Hoffmann (Amherst, NY: Prometheus Books, 2010), pp. 81–108, with corrections published in Richard Carrier, "Sources of the Jesus Tradition," May 30,

2011, at http://richardcarrier.blogspot.com/2011/05/sources-of-jesus-tradition.html, which should also be read with the associated adjunct document at http://www.richardcarrier .info/CarrierDec08.pdf, which concludes with a useful tutorial (on pp. 27–39, the section beginners will find the most useful after reading the present book). See also my evolving "Bayesian Calculator," at http://www.richardcarrier.info/bayescalculator.html. I also provide a discussion and application of BT in the endnotes and text of TCD_w as well as in TEC_s and TEC_r. For a historical background of the theory up to the present (including irrational hostility toward it and its eventual success in practice, see Sharon Bertsch McGrayne, *The Theory That Would Not Die: How Bayes' Rule Cracked the Enigma Code, Hunted Down Russian Submarines, and Emerged Triumphant from Two Centuries of Controversy* (New Haven, CT: Yale University Press, 2011).

7. For example, Richard Lempert, "Modeling Relevance," *Michigan Law Review* 75, no. 5/6 (April–May 1977): 1021–57; Daniel Kornstein, "A Bayesian Model of Harmless Error," *Journal of Legal Studies* 5, no. 1 (January 1976): 121–45. Semi-Bayesian methods have also been proposed in biblical textual analysis, though only at a very rudimentary level, for example, Winsome Munro, "Interpolation in the Epistles: Weighing Probability," *New Testament Studies* 36 (1990): 431–43; James Albertson, "An Application of Mathematical Probability to Manuscript Discoveries," *Journal of Biblical Literature* 78 (1959): 133–41.

8. The two best examples, which are designed as introductory texts on the subject and well recommended, are Caitlin Buck, William Cavanagh, and Clifford Litton, *Bayesian Approach to Interpreting Archaeological Data* (Chichester, England: Wiley, 1996); and W. G. Cavanagh et al., "Empirical Bayesian Methods for Archaeological Survey Data: An Application from the Mesa Verde Region," *American Antiquity* 72, no. 2 (April 2007): 241–72. Both demonstrate that BT methods are superior to subjective judgments, even by experts. And though both are very advanced, the general principles can still be grasped and employed without having to adopt the full apparatus of their methodology.

9. The best: E. T. Jaynes, *Probability Theory: The Logic of Science*, ed. G. Larry Bretthorst (Cambridge: Cambridge University Press, 2003); Luc Bovens and Stephan Hartmann, *Bayesian Epistemology* (Oxford: Oxford University Press, 2003); Richard Swinburne, ed., *Bayes's Theorem* (Oxford: Oxford University Press, 2002); and Earman, *Bayes or Bust?* (cited earlier in n. 1). Also of use is a Washington University website on Bayes's Theorem highlighting the work of E. T. Jaynes (bayes.wustl.edu) and the online Stanford Encyclopedia of Philosophy entries on Bayesian epistemology (plato.stanford .edu/entries/epistemology-bayesian) and Bayes's Theorem (plato.stanford.edu/entries/ bayes-theorem). A complete proof of the formal validity of BT in modern notation is provided in A. Papoulis, "Bayes' Theorem in Statistics" and "Bayes' Theorem in Statistics (Reexamined)," in *Probability, Random Variables, and Stochastic Processes*, 2nd ed. (New York: McGraw-Hill, 1984), pp. 38–39, pp. 78–81, and pp. 112–114 (cf. §3-5 and 4-4). For the most technical and advanced discussions: Peter Lee, *Bayesian Statistics: An*

Introduction, 3rd ed. (London: Hodder, 2004); James Berger, *Statistical Decision Theory and Bayesian Analysis*, 2nd ed. (New York: Springer-Verlag, 1993); Colin Howson and Peter Urbach, *Scientific Reasoning: The Bayesian Approach*, 2nd ed. (Chicago: Open Court, 1993); José Bernardo and Adrian Smith, *Bayesian Theory* (Chichester: Wiley, 1994); J. A. Hartigan, *Bayes Theory* (New York: Springer, 1983); Thomas Ferguson, *Mathematical Statistics: A Decision Theoretic Approach* (New York: Academic, 1967); and H. Jeffreys, *Theory of Probability*, 3rd ed. (Oxford: Oxford University Press, 1961).

10. Two notes for experts in advance of the ensuing discussion: (1) though it is common to omit conditioning information from notation if it runs through the whole analysis (i.e., the inclusion of *b* in the given equation for BT), this is only because it is already understood to be present, which is an assumption experts enjoy but novices do not, thus I will always include it as a reminder that it is indeed there and must always be included in one's understanding of what is being represented; and (2) scientists accustomed to employing BT may find my demarcation of *b* and *e* different from theirs; this difference reflects the different nature of history as a field of inquiry, whose data is more variegated and complex and almost always *post hoc*, which facts require the demarcation I employ (*b* as a typicality measure derived from the total field of background data, and *e* as the particular data to be explained by *h*), in contrast to what's more typical in science (*b* as all past data and *e* as all new data). Mathematically this makes no difference, as long as the demarcation is consistently applied (see chap. 6, p. 242).

11. The reason the actual frequency is still a factor is that dishonesty and error have a base rate in any given population, so as the frequency of the event being claimed decreases, so does the probability of the claim being true: $P(\text{TRUE CLAIM}|b) = N_1$ {the number of claimants telling the truth} / N_t {the total number of claimants}; $N_t = N_1$ {the number of claimants to whom the event actually happened} + N_2 {the number of claimants lying or in error}. As the frequency of the event decreases, so does N_1. But N_2 is not connected to the frequency of the event. It will stay more or less the same, or increase or decrease in various ways due to the influence of other factors (it can occasionally decrease in exactly the same proportion as N_1, but such a coincidence would have to be demonstrated and explained). So without any confirmed reason to expect N_2 to differ for this claim than for any other claim, a decrease in event frequency will decrease $P(\text{TRUE CLAIM}|b)$. But the $P(\text{TRUE CLAIM}|b)$ will still not *equal* the event frequency; it will instead equal N_1/N_t. This very same analysis entails that extremely *reliable* sources produce epistemic probabilities approaching the actual physical probabilities—as I will discuss in chapter 6 (p. 265).

12. The terms normally used for this (such as 'sampling distribution' and 'likelihood') are not well conceived in my opinion. This is a problem more obvious to laypeople, who are easily confused by these terms, than to mathematicians, who are so accustomed to them they are blind to their defects, e.g., in ordinary English (and thus to everyone in the humanities) "likelihood" is simply a synonym of "probability," and it would be a hard effort to

reprogram their brains to assign it so new and highly specialized a meaning as BT requires. A new convention is needed. A 'consequent' is the 'then' clause of a conditional proposition, and thus well suited here: $P(e)$ is the probability of e consequent to h; in other words, if h, then $P(e)$ (given b).

13. Instead, $P(\sim e|h.b)$, which is the probability of *not* having the evidence we do if h is true, and $P(e|h.b)$ must sum to 1 for any single h. In other words, it must always be the case that $P(e|h.b) + P(\sim e|h.b) = 1$ and $P(e|\sim h.b) + P(\sim e|\sim h.b) = 1$, therefore $P(e|h.b) = 1 - P(\sim e|h.b)$ and $P(e|\sim h.b) = 1 - P(\sim e|\sim h.b)$. I discuss how to make use of this fact in chapter 6 (p. 225).

14. I am assuming here of course that "supernatural" phenomena are not logically impossible, according to the definition I have developed elsewhere: Richard Carrier, "On Defining Naturalism as a Worldview," *Free Inquiry* 30, no. 3 (April/May 2010): 50–51; supported by Yonatan Fishman, "Can Science Test Supernatural Worldviews?" *Science & Education* 18, no. 6–7 (2007): 813–37.

15. Aptly noted by Alan Cromer in *Uncommon Sense: The Heretical Nature of Science* (New York: Oxford University Press, 1993). Of course, this was also a centerpiece of Karl Popper's philosophy of science, as described in *The Logic of Scientific Discovery* (New York: Basic Books, 1959) and updated (answering his critics) in *Conjectures and Refutations: The Growth of Scientific Knowledge*, 5th ed. (New York: Routledge, 1989). For the extent to which he was right, see Zuzana Parusniková and Robert S. Cohen, eds., *Rethinking Popper* (Dordrecht: Springer-Verlag, 2009).

16. The human brain has evolved many inept data-processing routines (cf. *TCD*, pp. 47–80 and *TEC*, pp. 155–78), and scientific methods are among the tools we developed to correct for them. But many of the brain's oldest (and most automatic and reliable) decision-making systems are already Bayesian: Laura Sanders, "The Probabilistic Mind: Human Brains Evolved to Deal with Doubt," *Science News* (October 8, 2011): 18–21.

17. Of course "historical facts" do *include* direct uninterpreted experience, because all observations of data and of logical and mathematical relations reduce to that, but no fact of history consists *solely* of that; and "the logically necessary and the logically impossible" *are* empirical facts in the trivial sense that they can be empirically observed, and empirical propositions depend on them, and logical facts are ultimately facts of the universe (in some fashion or other), but these are not empirical facts in the same sense as historical facts, because we cannot ascertain what happened in the past *solely* by ruminating on logical necessities or impossibilities. Logical facts are thus traditionally called *analytical* facts, in contrast to *empirical* facts. Some propositions might combine elements of both, but insofar as a proposition is at all empirical, it is not solely analytical (and thus has some nonzero epistemic probability of being true or false), and insofar as it is solely analytical, it is not relevantly empirical (and thus cannot affirm what happened in the past, but only what could or couldn't have).

18. We shouldn't be. Such attitudes are materially dangerous in modern society. See John Allen Paulos, *Innumeracy: Mathematical Illiteracy and Its Consequences* (New York: Hill and Wang, 1988) and Charles Seife, *Proofiness: The Dark Arts of Mathematical Deception* (New York: Viking, 2010). To get up to speed on all the basics of mathematics, there are two superb approaches designed for laypeople and liberal arts majors: Ronald Staszkow and Robert Bradshaw, *The Mathematical Palette*, 3rd ed. (Brooks Cole, 2004) and Marc Zev, Kevin Segal, and Nathan Levy, *101 Things Everyone Should Know about Math* (Washington, DC: Science, Naturally!, 2010). Danica McKellar (a beautiful actress who became a published mathematician in defiance of stereotype) has also begun a series of books to help the novice get the point that math is important and not as hard as your lousy high school teachers led you to believe: *Math Doesn't Suck: How to Survive Middle-School Math without Losing Your Mind or Breaking a Nail* (New York: Hudson Street Press, 2007), *Kiss My Math: Showing Pre-Algebra Who's Boss* (New York: Hudson Street Press, 2008), and *Hot X: Algebra Exposed* (New York: Hudson Street Press, 2010). Expect more to come. Though aimed at girls, they are just as useful to boys (being equally comprehensible and entertaining to either gender), and though marketed as being for 'middle schoolers' (through high school—the three books cover grades six through ten in the American school system), they are not too dumb for adults. When I was at a seminar on this topic (of applying math to history) an actual university professor asked me how you multiply percentages (because "100% × 80% would be 800% and what kind of result is that!?"); I then realized everyone should be reading McKellar. She is not a celebrity poser, by the way, but a real mathematician. The Chayes-McKellar-Winn Theorem is based on her published work: L. Chayes, D. McKellar, and B. Winn, "Percolation and Gibbs States Multiplicity for Ferromagnetic Ashkin-Teller Models on Z^2," *Journal of Physics A: Mathematical and General* 31, no. 45 (1998): 9055.

19. Some advanced Bayesian methods could still be applied to the study of history, but usually at very little relative gain at too enormous a cost in increased complexity—for example, Dempster-Shafer theory is specifically designed for knowledge fields like history where information is limited and ambiguous, but it's methodologically excruciating and requires extremely advanced mathematical skills. And yet, the outcome will rarely differ in any significant way, leaving its cost-benefit ratio for historians well beyond any benchmark of utility. I have already incorporated some elements of Dempster-Shafer theory in my presentation of Bayes's Theorem in this book, but in a fashion that bypasses its extremely complex apparatus. For those nevertheless still interested see Arthur Dempster, "A Generalization of Bayesian Inference," *Journal of the Royal Statistical Society* 30, Series B (1968): 205–47; Glenn Shafer, *A Mathematical Theory of Evidence* (Princeton, NJ: Princeton University Press, 1976); Kari Sentz and Scott Ferson, "Combination of Evidence in Dempster-Shafer Theory," technical report, Sandia National Laboratories SAND 2002 -0835, available at http://www.sandia.gov/epistemic/Reports/SAND2002-0835.pdf.

20. Which even scientists can be guilty of: Tom Siegfried, "Odds Are, It's Wrong: Science

Fails to Face the Shortcomings of Statistics," *Science News* 177, no. 7 (March 27, 2010): 26–29 (with supplemental materials at http://bit.ly/aq1x28; see also the more classic analysis of Jerome Cornfield, "The Frequency Theory of Probability, Bayes' Theorem, and Sequential Clinical Trials," in *Bayesian Statistics*, Donald Meyer and Raymond Collier Jr., ed. [Itasca, IL: F. E. Peacock, 1968], pp. 1–28). Notably, Siegfried argues that better acquaintance with Bayes's Theorem may be the solution, citing a detailed defense of this fact by George Diamond and Sanjay Kaul, "Prior Convictions: Bayesian Approaches to the Analysis and Interpretation of Clinical Megatrials," *Journal of the American College of Cardiology* 43, no. 11 (June 2, 2004): 1929–39, which notably describes a formal model for *a fortiori* reasoning with Bayes's Theorem (a crucial idea I will discuss later in this chapter, albeit less formally) and argues that such a method should become an empirical standard. I agree.

21. You should also read Kees van Deemter, *Not Exactly: In Praise of Vagueness* (Oxford: Oxford University Press, 2010), which provides a thorough and convincing defense of the necessity and logical validity of this and many other kinds of imprecision.

22. On the validity of this conclusion, see Sérgio B. Volchan, "Probability as Typicality," *Studies in History and Philosophy of Science Part B: Studies in History and Philosophy of Modern Physics* 38, no. 4 (December 2007): 801–14. I discuss this point again in chapter 6 (p. 250).

23. See my demonstration of this in TCD_w, pp. 298–99 (with the corresponding Bayesian notation given in the notes on pp. 310–11). I also discuss the logical validity of that principle here in chapter 4 (p. 114), and its application in chapter 5 (p. 177).

24. Yudkowsky (in "An Intuitive Explanation") discusses examples where even expert medical scientists have made hugely incorrect estimates of probability that can (and actually have) lead to wasteful and even harmful medical decisions, all because the actual logic of evidence is not as simple as even experts regularly assume. The recent confusion over mammograms and breast cancer overtreatment is one such case (see McGrayne, *The Theory That Would Not Die*, pp. 255–58).

25. See *NIF* and TEC_s.

26. Several chapters in *TCD* support this conclusion.

27. Because as stipulated A = (D and C), and B = (D and ~C), and P(D) = P(D and C) + P(D and ~C).

28. Note that the remainder of this chapter (and much of chapter 6) addresses all the concerns regarding BT raised in C. Behan McCullagh, *Justifying Historical Descriptions* (New York: Cambridge University Press, 1984), p. 58.

29. The following points are smartly supported by Giulio D'Agostini, "Role and Meaning of Subjective Probability: Some Comments on Common Misconceptions" in Ali Mohammad-Djafari, ed., *Bayesian Inference and Maximum Entropy Methods in Science and Engineering* (Melville, NY: American Institute of Physics, 2001): 23–30, also available at http://arxiv.org/abs/physics/0010064.

30. For a complete demonstration of this point (and supporting my arguments to follow), see Giulio D'Agostini, "Teaching Statistics in the Physics Curriculum: Unifying and Clarifying Role of Subjective Probability," *American Journal of Physics* 67, no. 12 (December 1999): 1260–68.

31. It's well worth reading the astute remarks on this point by Sandra LaFave, "Thinking Critically about the 'Subjective'/'Objective' Distinction," http://instruct.west valley.edu/lafave/subjective_objective.html.

32. This happens *a lot* in history, an outcome many scientists will find unfamiliar. Imagine a drug study for treating the common cold in which all five of the first patients in the cohort immediately die as soon as they receive the treatment; the study would be canceled at once, no further collection of data required. Though it's statistically possible those deaths were a fluke, the odds are already too small to credit that hypothesis, even with a sample of only five. Analogously, once we find five diverse and independent cases of silent reading in the extant evidence from antiquity, we don't have to bother collecting more data to conclude 'ancient readers could and sometimes did read silently' (see n. 10 for chap. 2, pp. 298–99).

33. The objection might be raised that Bayesian argumentation could be used to "hoodwink" people by cloaking bad arguments in the regalia of mathematical language that nonexperts cannot evaluate. But this objection applies to all methods whatever, even the semantics of ordinary language already employed by historians (as more than amply demonstrated in Hackett Fischer, *Historians' Fallacies*), but especially standard scientific statistical languages (many a complex statistical argument has been used to hoodwink the public) and symbolic logic (many a philosophy paper has disguised bad reasoning in formalized languages impenetrable to nonexperts). The solution is obviously not to abandon such methods, but to master them and police their abuse. As for symbolic logic, classical statistics, and ordinary language, so for BT.

34. You should also be familiar with the following symbols: = (equals); > (greater than); < (less than); >> (much greater than); << (much less than); >>> (very much greater than); <<< (very much less than); ≥ (greater than or equal to); ≤ (less than or equal to); ≈ (approximately equal to); and → (approaching); the latter symbol will also mean other things in different contexts: in contexts pertaining to causal sequence it will mean 'and then'; and, pertaining to logical entailment, 'entails.'

CHAPTER 4. BAYESIAN ANALYSIS OF HISTORICAL METHODS

1. John Earman, *Bayes or Bust? A Critical Examination of Bayesian Confirmation Theory* (Cambridge, MA: MIT Press, 1992), p. 63. I discuss some of Earman's pro-and-con analysis in chapter 6.

2. See *SGG*, pp. 227–52.

3. As can be inferred from all the leading works in historical method. Most directly: C. Behan McCullagh, *The Logic of History: Putting Postmodernism in Perspective* (New York: Routledge, 2004), *The Truth of History* (New York: Routledge, 1998), and *Justifying Historical Descriptions*. Most recently: David Henige, *Historical Evidence and Argument* (Madison: University of Wisconsin Press, 2005); Marc Trachtenberg, *The Craft of International History: A Guide to Method* (Princeton, NJ: Princeton University Press, 2006); J. Tosh, *The Pursuit of History: Aims, Methods and Directions in the Study of Modern History*, 4th ed. (New York: Longman, 2006); and Allan Megill, *Historical Knowledge, Historical Error: A Contemporary Guide to Practice* (Chicago: University of Chicago Press, 2007). Likewise: Walter Prevenier and Martha Howell, *From Reliable Sources: An Introduction to Historical Methods* (Ithaca, NY: Cornell University Press, 2001); Clayton Roberts, *The Logic of Historical Explanation* (University Park: Pennsylvania State University Press, 1996); Joyce Appleby, Lynn Hunt, and Margaret Jacob, *Telling the Truth about History* (New York: Norton, 1995); Robert Shafer, *A Guide to Historical Method*, 3rd ed. (Homewood, IL: Dorsey, 1980); David Hackett Fischer, *Historians' Fallacies: Toward a Logic of Historical Thought* (New York: Harper & Row, 1970); Homer Hockett, *The Critical Method in Historical Research and Writing* (New York: Macmillan, 1955), see especially part 1; Louis Gottschalk, *Understanding History: A Primer of Historical Method* (New York: Knopf, 1950); and Gilbert Garraghan, *A Guide to Historical Method* (New York: Fordham University Press, 1946). And reinforcing those (though more from the perspective of epistemology than method): John Lewis Gaddis, *The Landscape of History: How Historians Map the Past* (New York: Oxford University Press, 2002); G. R. Elton, *The Practice of History*, 2nd ed. (New York: Wiley-Blackwell, 2002); David Cannadine, ed., *What Is History Now?* (New York: Palgrave Macmillan, 2002); Richard Evans, *In Defence of History* (London: Granta Books, 1997) and *Return to Essentials: Some Reflections on the Present State of Historical Study* (New York: Cambridge University Press, 1991); Robin Collingwood, *The Idea of History (with Lectures 1926–1928)* (New York: Oxford University Press, 1994) and *The Principles of History and Other Writings in Philosophy of History* (New York: Oxford University Press, 1999); Murray Murphey, *Philosophical Foundations of Historical Knowledge* (Albany: State University of New York Press, 1994); J. L. Gorman, *Understanding History: An Introduction to Analytical Philosophy of History* (Ottawa, ON: University of Ottawa Press, 1992); Arthur Danto, *Narration and Knowledge (Including the Integral Text of Analytical Philosophy of History)* (New York: Columbia University Press, 1985); Paul Veyne, *Writing History: Essay on Epistemology* (Middletown, CT: Wesleyan University Press, 1984); John Lukacs, *Historical Consciousness: The Remembered Past* (New York: Harper & Row, 1968) and *The Future of History* (New Haven, CT: Yale University Press, 2011); Morton Gabriel White, *Foundations of Historical Knowledge* (New York, Harper & Row, 1965); William Dray, ed., *Philosophical Analysis*

and History (New York: Harper & Row, 1966); Edward Carr, *What Is History?* (New York: Knopf, 1961); and Marc Bloch, *The Historian's Craft* (New York: Knopf, 1953).

4. Note that this category of evidence (the unbiased or counterbiased source) is related to the Argument from Embarrassment, which I will examine in considerable detail in the next chapter (starting on p. 124, there from the different perspective of *friendly* sources rather than neutral or hostile).

5. I describe and discuss this miracle and cite the scholarship on it in *SGG*, pp. 228–31.

6. C. Behan McCullagh, *Justifying Historical Descriptions* (New York: Cambridge University Press, 1984), p. 19. That all "inferences to the best explanation" are inherently Bayesian is argued in Peter Lipton, *Inference to the Best Explanation*, 2nd ed. (Routledge, 2004); on which see *Notre Dame Philosophical Reviews*, June 1, 2005, http://ndpr.nd.edu/news/24796-inference-to-the-best-explanation-2nd-edition.

7. Formally demonstrated by William H. Jefferys and James O. Berger, "Sharpening Ockham's Razor on a Bayesian Strop," Technical Report #91-44 C, Department of Statistics, Purdue University, August 1991, available at http://bayesrules.net/papers/ockham.pdf. See also E. T. Jaynes, *Probability Theory: The Logic of Science*, ed. G. Larry Bretthorst (Cambridge: Cambridge University Press, 2003), pp. 601–13; and P. M. B. Vitanyi and Ming Li, "Minimum Description Length Induction, Bayesianism, and Kolmogorov Complexity," *IEEE Transactions on Information Theory* 46, no. 2 (March 2000): 446–64.

8. I suspect prior probability is also implicit in standard statistical significance tests, even when mathematicians don't realize it, and that in result significance testing is also reducible to BT. To say the RESULT of a statistical test has a *p*-value of 5 percent is to say that this RESULT is only 5 percent likely to have been produced by chance, which in Bayesian terms is equivalent to assuming that the prior probability that chance caused *e* is 0.05 and that *h* is equivalent to "chance did not cause *e*" and ~*h* to "chance *did* cause *e*." Because "chance did not cause *e*" is what it formally means to reject the null hypothesis. To say "either the null hypothesis is false *or* an unusual event has occurred" is to say that, given the data observed, the null hypothesis is true only if the result was produced by chance and not by the hypothesized phenomenon (the HYPOTHESIS), which is simply to say that the null hypothesis is that chance produced the observed result, which in Bayesian terms is ~*h*. It is then being covertly assumed that all other hypotheses (e.g., fraud, magic, other uncontrolled variables) that are not being tested all have a prior probability of zero (or as near to zero as can be ignored), based on the not-always-valid assumption that experimental controls, like double-blind procedures, have eliminated them (in reality, such procedures can at best only *reduce the probability* of other causes, not eliminate them). (I'm not the first to suggest this; see the articles by Siegfried and by Diamond and Kaul cited in n. 20 for chap. 3, pp. 303–304.) One might think converting a standard test into Bayesian terms would make the *p*-value the *consequent* probability of the evidence on a hypothesis of chance, but in fact the claim being made by significance testing is that, for a *p*-value of 5 percent, $P(\text{CHANCE}|\text{DATA}) = 0.05$,

which is P(~h|e), a posterior probability, which in BT is entailed by P(CHANCE) = 0.05 (a 0.05 prior probability) and P(RESULT|CHANCE) = 1 (a consequent probability of 1, the same value assumed for P(RESULT|HYPOTHESIS)). This is because the *p*-value represents the probability of a comparable result *when the null hypothesis is true*, and the null hypothesis is mathematically equivalent to the hypothesis that random chance caused the result, so assuming that that hypothesis is *true*, i.e., that chance *did* cause *e*, is in effect to assign chance a consequent probability of 1, leaving us only the task of ascertaining how often chance would do that, which happens to be the *p*-value. In other words, to declare a *p*-value of 0.05 is to declare that 1 in 20 times chance will be the cause of data like the RESULT, and if 1 in 20 times chance will do that, then the prior probability that chance will produce something like the RESULT is 0.05 and the RESULT will be observed in 100 percent of those 1 in 20 times (every time both the null hypothesis is true and that result is observed). See my analogous discussion of determining the prior probability of being cheated at poker in chapter 6 (p. 254).

9. Demonstrated in detail by Jaynes and Bretthorst, *Probability Theory*.

10. For some examples, see my essay on "Experimental History," July 28, 2007, at http://richardcarrier.blogspot.com/2007/07/experimental-history.html.

11. P3 = ~(B and ~A), from which it follows by *modus tollens*, if B, then ~(~A); ergo, if B, then A; P4 = ~(A and C), from which follows by *modus tollens*, if B, then ~C; if we accept logic, then B; ergo, ~C.

12. See Tom Siegfried, "Odds Are, It's Wrong: Science Fails to Face the Shortcomings of Statistics," *Science News* 177, no. 7 (March 27, 2010) and George Diamond and Sanjay Kaul, "Prior Convictions: Bayesian Approaches to the Analysis and Interpretation of Megatrends," *Journal of the Amerian College of Cardiology* 43, no. 11 (June 2, 2004). See n. 20 for chap. 3 (pp. 303–304). I suspect we could go even further (see n. 8 above, and chap. 6, p. 265).

13. By disjunction: either A (the priors are equal) or B (they are not); therefore, if ~A, then necessarily B; and if B, then BT will entail a different probability than any method that ignores a difference in priors; therefore any such method will contradict BT. In other words, if the priors *are* different, then you *must* account for that in your math, otherwise your result will not be valid, logically or mathematically.

14. The underlying logic shouldn't have to be laid out, but here it is:

1. Either S (a method says the same things about the epistemic probability of *h* as BT) or L (a method says less than BT) or M (a method says more than BT).
2. If S, then either A (the method produces the same conclusion from those same premises) or C (it produces a different conclusion from those same premises).
3. If A, then that method is reducible to BT (and is therefore D), and if C, then that method is invalid (per C2). So if S, then either D or C; ~C (per C2); ergo, if S, then D.

4. If L, then either F (the method assumes all the same premises as BT without stating or verifying them all) or E (it omits premises that are otherwise known to validly affect the probability of the claim being true).

5. E is invalid (by definition a non sequitur: an argument whose conclusion is dependent on some known fact K cannot produce a valid conclusion without accounting for K); ergo if E, then C.

6. So if F, then S; and if S, then D (per above); and if E, then C; ~C (per C2); ergo, ~E; so if L, then either D or E; ~E; ergo, if L, then D.

7. And if P5 and P6 are true, then ~M, and if ~M, then either S or L; but (as just shown), if S or L, then D; ergo, D.

8. Therefore, if P5 and P6, then only D.

Note that P6 is true by definition; therefore, the burden is on anyone to challenge P5 with a valid counterexample. Otherwise, the conclusion follows.

15. See further discussion in TCD_w and SGG (especially pp. 209–52, with pp. 154–57, pp. 159–60). Hundreds of examples are cataloged in the books and articles of Joe Nickel, James Randi, Robert Todd Carroll, and Massimo Polidoro; the websites of the Committee for Skeptical Inquiry (CSI) (http://www.csicop.org) and Snopes (http://www.snopes.com); in various anthologies and issues of the *Skeptic* and the *Skeptical Inquirer*; and in various encyclopedias, including *The Skeptic Encyclopedia of Pseudoscience* (Santa Barbara, CA: ABC-CLIO, 2002) and *An Encyclopedia of Claims, Frauds, and Hoaxes of the Occult and Supernatural* (New York: St. Martin's, 1995). For an essential bibliography and discussion on classifying and explaining "miracles" in historical and oral reports, see Robert Shanafelt, "Magic, Miracle, and Marvels in Anthropology," *Ethnos* 69, no. 3 (September 2004): 317–40; with support (especially in the abundant scholarship cited) in Craig Keener, "A Reassessment of Hume's Case against Miracles in Light of Testimony from the Majority World Today," *Perspectives in Religious Studies* 38, no. 3 (Fall 2011): 289–310.

16. See the extensive collection of examples in Stith Thompson, *Motif-Index of Folk-Literature: A Classification of Narrative Elements in Folk-Tales, Ballads, Myths, Fables, Mediaeval Romances, Exempla, Fabliaux, Jest-books, and Local Legends*, vols. 1–6 (Helsinki: Academia Scientiarum Fennica, 1932–1936); Louis Ginzberg, *The Legends of the Jews*, vols. 1–7 (Philadelphia: Jewish Publication Society of America, 1909–1938); and Wendy Cotter, *Miracles in Greco-Roman Antiquity: A Sourcebook for the Study of New Testament Miracle Stories* (New York: Routledge, 1999).

17. The pervasive unreliability of ancient historians and biographers is well proven by now, varying only in degree (but notoriously at or near its worst in the case of hagiography, i.e., the biographies of heroes and saints). See, for example, Charles Fornara, *The Nature of History in Ancient Greece and Rome* (Berkeley: University of California Press, 1983); Michael Grant, *Greek and Roman Historians: Information and Misinformation* (London:

Routledge, 1995); A. B. Bosworth, *From Arrian to Alexander: Studies in Historical Interpretation* (Oxford: Oxford University Press, 1988); Alan Cameron, *Greek Mythography in the Roman World* (New York: Oxford University Press, 2004); Mary Lefkowitz, *The Lives of the Greek Poets* (Baltimore: Johns Hopkins University Press, 1981); and Ava Chitwood, *Death by Philosophy: The Biographical Tradition in the Life and Death of the Archaic Philosophers Empedocles, Heraclitus, and Democritus* (Ann Arbor: University of Michigan Press, 2004). See also *TET$_s$*, pp. 168–82 and *TCD$_w$*, pp. 291–93.

18. Accordingly, Hume's antiquated argument against miracles has been corrected using BT, verifying my conclusion here: Aviezer Tucker, "Miracles, Historical Testimonies, and Probabilities," *History and Theory* 44 (October 2005): 373–90 (with further sources cited, e.g., p. 374, n. 3); Robert Fogelin, *A Defense of Hume on Miracles* (Princeton, NJ: Princeton University Press, 2003); Michael Levine, "Bayesian Analyses of Hume's Argument Concerning Miracles," *Philosophy and Theology* 10, no. 1 (1997): 101–106; Jordan Howard Sobel, "On the Evidence of Testimony for Miracles: A Bayesian Interpretation of David Hume's Analysis," *Philosophical Quarterly* 37, no. 147 (April 1987): 166–86. See also Mark Strauss, *Four Portraits, One Jesus: An Introducrtion to Jesus and the Gospels* (Grand Rapids, MI: Zondervan, 2007), pp. 456–68 (with pp. 363–65); and Yonatan Fishman, "Can Science Test Supernatural Worldviews?" *Science and Education* 18, no. 6–7 (August 2007): 813–37. Also pertinent is Jaynes's Bayesian treatment of ESP, in Jaynes and Bretthorst, *Probability Theory*, pp. 119–48. As a result, while "naive" Humean arguments against miracles are soundly refuted in Keener ("A Reassessment"), sound Bayesian reconstructions (such as I have briefed here) are not.

19. See, for example, Israel Finkelstein and Neil Silberman, *The Bible Unearthed: Archaeology's New Vision of Ancient Israel and the Origin of Its Sacred Texts* (New York: Free Press, 2001) and Thomas Thompson, *The Mythic Past: Biblical Archaeology and the Myth of Israel* (New York: Basic Books, 1999). Scholars all across the spectrum agree on this point (apart from biblical literalists, who prefer dogma to logically valid empirical argument), e.g., William Dever, *What Did the Biblical Writers Know, and When Did They Know It? What Archaeology Can Tell Us about the Reality of Ancient Israel* (Grand Rapids, MI: William B. Eerdmans, 2001) on the one hand, and Avalos, *The End of Biblical Studies* on the other.

20. Garraghan, *Guide to Historical Method*, §149a. See also Shafer, *Guide to Historical Method*, p. 77; Gottschalk, *Understanding History*, pp. 45–46; and Neville Morley, *Writing Ancient History* (Ithaca, NY: Cornell University Press, 1999), pp. 67–68.

CHAPTER 5. BAYESIAN ANALYSIS OF HISTORICITY CRITERIA

1. Christopher Tuckett, "Sources and Methods," in *The Cambridge Companion to Jesus*, ed. Markus Bockmuehl (Cambridge: Cambridge University Press, 2001), pp. 132–33. That this seems to contradict the more usual rule of rejecting inaccuracies and anachronisms in a story is a point I'll return to (on p. 176).

2. Mark Allan Powell, "Sources and Criteria," in *Jesus as a Figure in History: How Modern Historians View the Man from Galilee* (Louisville, KY: John Knox, 1998), pp. 31–50, 187–89 (quoting p. 47).

3. The example given is that 'Jesus intimately addressing God as his father is dissimilar to Jewish practice,' which is false: see Mary Rose D'Angelo, "Abba and Father: Imperial Theology in the Contexts of Jesus and the Gospels," in *The Historical Jesus in Context*, ed. Amy-Jill Levine, Dale C. Allison Jr., and John Dominic Crossan (Princeton, NJ: Princeton University Press, 2006), pp. 64–78; and Joachim Jeremias, "*Abba* as an Address to God," in *The Historical Jesus in Recent Research*, ed. James Dunn and Scot McKnight (Winona Lake, IN: Eisenbrauns, 2005), pp. 201–206.

4. Tuckett, "Sources and Methods," p. 132.

5. See Stanley Porter, *The Criteria for Authenticity in Historical-Jesus Research: Previous Discussion and New Proposals* (Sheffield, UK: Sheffield Academic Press, 2000), pp. 70–76, pp. 114–16; and Gerd Theissen and Dagmar Winter, *The Quest for the Plausible Jesus: The Question of Criteria*, trans. M. Eugene Boring (Louisville, KY: John Knox Press, 2002), which is entirely devoted to refuting the validity of this single criterion. I summarize more arguments and examples against its validity in Richard Carrier, "Bayes' Theorem for Beginners: Formal Logic and Its Relevance to Historial Method," in Sources of the Jesus Tradition: Separating History from Myth, ed. R. Joseph Hoffmann (Amherst, NY: Prometheus Books, 2010), pp. 90–92.

6. Dennis Ingolfsland, "The Historical Jesus according to John Dominic Crossan's First Strata Sources: A Critical Comment," *Journal of the Evangelical Theological Society* 45, no. 3 (2002): 405–414 (quoting p. 413, n. 40).

7. Theissen and Winter, *Quest for the Plausible Jesus*, p. 168.

8. For the most useful discussion, including its origins and historical development: Porter, *Criteria for Authenticity*, pp. 106–10 (its history: pp. 106–107, esp. note 9). John Meier, "Criteria: How Do We Decide What Comes from Jesus?" *A Marginal Jew: Rethinking the Historical Jesus*, vol. 1 (New York: Doubleday, 1991), pp. 168–71, provides the most well-known discussion (mostly credulous, but with some critique). That the EC is a variant of the more general criterion of dissimilarity (which entails asking more generally "why would Christians attribute that statement or behavior to Jesus unless it were true?"), see Dennis Polkow, "Method and Criteria for Historical Jesus Research," in *Society of*

Biblical Literature 1987 Seminar Papers, ed. Kent Harold Richards (Atlanta, GA: Scholars Press, 1987), pp. 336–56 (cf. p. 341).

9. For example: Bernard Jefferson, "Declarations against Interest: An Exception to the Hearsay Rule," *Harvard Law Review* 58, no. 1 (Noveber 1944): 1–69; John Capowski, "Statements against Interest, Reliability, and the Confrontation Clause," *Seton Hall Law Review* 28 (1997): 471–511. Examples of the current rule at law are Rule 804(b)(3) of the US Federal Rules of Evidence and Section 230 of the California Evidence Code.

10. Meier, *Marginal Jew*, vol. 1, p. 168.

11. Mark's penchant for fabrication is often denied, but is fairly well confirmed by now, e.g., Randel Helms, *Gospel Fictions* (Amherst, NY: Prometheus Books, 1988); Burton Mack, *A Myth of Innocence: Mark and Christian Origins* (Philadelphia: Fortress Press, 1988); Dennis MacDonald, *The Homeric Epics and the Gospel of Mark* (New Haven, CT: Yale University Press, 2000); Thomas Thompson, *The Messiah Myth: The Near Eastern Roots of Jesus and David* (New York: Basic Books, 2005). See also Michael Vines, *The Problem of Markan Genre: The Gospel of Mark and the Jewish Novel* (Atlanta, GA: Society of Biblical Literature, 2002). On this same subject I have also found useful the cautious but often apt analysis of R. G. Price, *The Gospel of Mark as Reaction and Allegory* (Raleigh, NC: Lulu.com, 2007). I will demonstrate this point in my next volume, *On the Historicity of Jesus Christ*.

12. As catalogued, for example, in Wayne Kannaday, *Apologetic Discourse and the Scribal Tradition: Evidence of the Influence of Apologetic Interests on the Text of the Canonical Gospels* (Atlanta, GA: Society of Biblical Literature); Bart Ehrman, *The Orthodox Corruption of Scripture: The Effect of Early Christological Controversies on the Text of the New Testament* (New York: Oxford University Press, 1993); and C. S. C. Williams, *Alterations to the Text of the Synoptic Gospels and Acts* (Oxford, UK: Basil Blackwell, 1951).

13. For a full discussion, see *TET*, pp. 358–64.

14. Porter, *Criteria for Authenticity*, p. 109 (see pp. 106–110 for Porter's full critique of the EC).

15. Mark Strauss, *Four Portraits, One Jesus: An Introduction to Jesus and the Gospels* (Grand Rapids, MI: Zondervan, 2007), p. 361. Meier concedes the same point: Meier, *Marginal Jew*, vol. 1, p. 170.

16. Theissen and Winter, *Quest for the Plausible Jesus*, p. 175.

17. M.D. Hooker, "Christology and Methodology," *New Testament Studies* 17 (1970): 482.

18. I demonstrate this point with numerous examples, and explore the sociological and anthropological underpinnings of it, in *NIF*.

19. Hooker, "Christology and Methodology," p. 482.

20. For example, we know almost nothing of the context behind 1 Corinthians 15:29, or the whole of 1 Corinthians 12 and 14, or the myriad undescribed "other gospels" and "other communities" of Christians competing with Paul's (e.g., Galatians 1:6–9; 1 Corinthians

1:12, 3:4–6; 2 Corinthians 11:4, 13; Romans 16:17–18; Philemon 1:15–17; and after Paul: 2 Thessalonians 2:2–5, 15; 1 Timothy 4:1–3, 7, 5:15; 2 Timothy 2:16–18, 3:4–7, 9–10, 13–14; 2 Peter 2:1–3, 3:16; 1 John 4:1; Jude 3–4, 8–16; Hebrews 13:8–9). The diversity of Jewish thought in the same period is likewise bewildering, and our ignorance extends well beyond even that: *TET*ₛ, pp. 107–10, 180–82.

21. Craig Evans, "Life-of-Jesus Research and the Eclipse of Mythology," *Theological Studies* 54 (1993): 3–36 (referencing pp. 24–26).

22. Meier, *Marginal Jew*, vol. 1, p. 170.

23. Ibid., pp. 170–71. Meier's point is not of his own invention; it had already been thoroughly confirmed by previous scholars (see n. 43 below, p. 315). On Mark's extensive use of the Psalms here, see Douglas Moo, *The Old Testament in the Gospel Passion Narratives* (Sheffield, UK: Almond Press, 1983), pp. 264–83.

24. I demonstrate this in *TET*ₛ, pp. 163–65, but it's also argued in Jerry Camery-Hoggatt, *Irony in Mark's Gospel: Text and Subtext* (New York: Cambridge University, 1992), and many other scholars have remarked upon it. For example: Paul Danove, *The End of Mark's Story: A Methodological Study* (Leiden, Netherlands: Brill, 1993) and Adela Collins, "The Empty Tomb in the Gospel According to Mark," in *Hermes and Athena: Biblical Exegis and Philosophical Theology*, ed. Eleonore Stump and Thomas Flint (Notre Dame, IN: University of Notre Dame, 1993), pp. 107–40.

25. For a full discussion of the evidence, see *TET*ₛ, pp. 158–61.

26. For the most concise expert demonstration of this point, see Gerd Lüdemann, "Paul as a Witness to the Historical Jesus," in *Sources of the Jesus Tradition: Separating History from Myth*, ed. Joseph Hoffmann (Amherst, NY: Prometheus, 2010), pp. 196–212. Note that many of the New Testament Epistles are now generally recognized as forgeries written a century later (and hence must be excluded from this category of evidence). For a good summary of the evidence and scholarship on this point, see Bart Ehrman, *Forged: Writing in the Name of God—Why the Bible's Authors Are Not Who We Think They Are* (New York: HarperOne, 2011).

27. Evans, "Life-of-Jesus Research," pp. 24–26.

28. For both examples, see *NIF*, pp. 17–49.

29. See Timothy Peter Wiseman, *The Myths of Rome* (Exeter, UK: University of Exeter Press, 2004), pp. 46–47, 138–48.

30. Of course we now know that entire prophecy was a fabrication, the book of Daniel having been forged centuries after it purports to have been written, cf., e.g., André Lacocque, *The Book of Daniel* (Louisville, KY: John Knox, 1979). Although some early pagan critics of Christianity had noticed this, too. On this and other evidence of Jewish speculation regarding a "dying messiah" who would redeem Israel or otherwise presage the end of the world, see *NIF*, pp. 34–44 and Richard Carrier, "The Dying Messiah," October 5, 2011, http://richardcarrier.blogspot.com/2011/10/dying-messiah.html. Key evidence includes a

pre-Christian Jewish pesher recovered from Qumran that makes this claim explicit, and at the same time links the dying messiah of Daniel 9 to the suffering servant of Isaiah 52–53 (11QMelch ii.18, aka 11Q13, http://www.gnosis.org/library/commelc.htm).

31. Mark 13:14 quotes Daniel directly (and Matthew provides the attribution: Matthew 24:15); on Matthew's allusions to the tale of Daniel in the lion's den, see later in this chapter (p. 199). Matthew connects Daniel to the nativity by including *magi*, a term that appears nowhere else in the Bible except in Daniel (in the Greek), and paralleling and reversing elements therein (e.g., in Daniel, kings are troubled by omens and summon their wise men to explain them, including the magi and a foreigner, Daniel; in Matthew, the king is troubled by an omen and summons his wise men to explain it, including the magi, who this time are the foreigners).

32. Porter, *Criteria for Authenticity*, p. 110.

33. As noted earlier, Porter, *Criteria for Authenticity*, and Theissen and Winter, *Quest for the Plausible Jesus*, are effectively devoted to demonstrating this.

34. Quoting Goodacre's weblog comments (from November 21, 2005) now archived at http://web.archive.org/web/20080921090341/http://ntgateway.com/weblog/2005/11/sbl -monday-afternoon.html, which he has developed into a detailed argument against both the EC and multiple attestation criteria in Mark Goodacre, "Criticizing the Criterion of Multiple Attestation: The Historical Jesus and the Question of Sources," in *Jesus, History and the Demise of Authenticity*, ed. Chris Keith and Anthony LeDonne (New York: T & T Clark, forthcoming, 2012). See also Rafael Rodriguez, "The Embarrassing Truth about Jesus: The Demise of the Criterion of Embarrassment," in the same volume.

35. John Meier, "The Circle of the Twelve: Did It Exist During Jesus' Ministry?" *Journal of Biblical Literature* 116, no. 4 (Winter 1997): 665 [635–72].

36. I discuss both these facts (the likely influence of pagan and Old Testament precedents on the core structure of the Christian Gospel) in more detail in *NIF*, pp. 17–49; see note 30 above (p. 313). I will cover the evidence in even greater detail in my next volume, *On the Historicity of Jesus Christ*.

37. This is essentially the argument of Earl Doherty, *The Jesus Puzzle: Did Christianity Begin with a Mythical Christ?* (Ottawa, ON: Canadian Humanist, 1999) and *Jesus: Neither God nor Man* (Ottawa, ON: Age of Reason, 2009); the argument is not as far-fetched as usually assumed, but I will examine it properly in the next volume.

38. The evidence at Qumran is cited in note 30 (p. 313).

39. See TET_b, pp. 373–79.

40. According to the Talmud, *b.Sanhedrin* 43a.

41. As the Gospels could seem to imply, suggesting Jesus may have declared himself king and marshaled an armed force: e.g., Mark 14:47, John 18:10, Luke 22:36–38, and all of Mark 15 and parallels. For the most valiant attempt to make this case, see S. G. F. Brandon, *Jesus and the Zealots: A Study of the Political Factor in Primitive Christianity* (Manchester,

UK: Manchester University Press, 1967). But it would not take much for the Romans to mis-understand apocalyptic religious talk as code for armed rebellion and convict and execute Jesus on those grounds alone, as demonstrated by their equally decisive response to other occasions of that very thing illustrated in Josephus, cf. Craig Evans, "Josephus on John the Baptist and Other Jewish Prophets of Deliverance," in Levine, Allison, and Crossan, *Historical Jesus in Context*, pp. 55–63. The later execution of Christians for arson may reflect exactly the same misunderstanding: just read Tacitus, *Annals* 15.44 in light of 2 Peter 3.

42. See *TET*_b, pp. 375–77.

43. Which was Meier's own point about the cry of dereliction; this has been thor-oughly confirmed by other scholars: George Nickelsburg, "*First* and *Second Enoch*: A Cry against Oppression and the Promise of Deliverance," in Levine, Allison, and Crossan, *Historical Jesus in Context*, pp. 87–109, and "The Genre and Function of the Markan Passion Narrative," *Harvard Theological Review* 73, no. 1/2 (January–April 1980): 153–84; see also Thompson, *Messiah Myth*, pp. 191–93.

44. See *NIF*, pp. 51–83.

45. Eric Laupot, "Tacitus' Fragment 2: The Anti-Roman Movement of the *Christiani* and the *Nazoreans*," *Vigiliae Christianae* 54, no. 3 (2000): 233–47.

46. J. S. Kennard, "Was Capernaum the Home of Jesus?" *Journal of Biblical Literature* 65, no. 2 (June 1946): 131–41; and "Nazorean and Nazareth," *Journal of Biblical Literature* 66, no. 1 (March 1947): 79–81, responding to W. F. Albright's reply in "The Names Nazareth and Nazoraean," *Journal of Biblical Literature* 65, no. 4 (December 1946): 397–401.

47. Gospel of Phillip 66:14, 56:12, 62:8, 62:15.

48. René Salm, *The Myth of Nazareth: The Invented Town of Jesus* (Cranford, NJ: American Atheist, 2008): pp. xii–xiii and 299, n. 109(c).

49. Irenaeus, *Against All Heresies* 1.21.3. We do not know of any such word; more likely the actual derivation was from something else, like *natsar*, as perhaps "keeper of secrets" (i.e., the mysteries; by derivation from, e.g., Isaiah 48:6 and 42:6), which Christians proclaimed (and thus equated with) "the truth."

50. See, e.g., Susan Levin, "Platonic Eponymy and the Literary Tradition," *Phoenix* 50, no. 3/4 (Autumn–Winter 1996): 197–207.

51. See Kennard in note 46 above.

52. Salm, *Myth of Nazareth*, pp. 299–305, is admittedly right about this. The matter is also well discussed in Robert Price, *The Incredible Shrinking Son of Man: How Reliable Is the Gospel Tradition?* (Amherst, NY: Prometheus Books, 2000), pp. 51–54.

53. Note that Matthew is known to provide (or correct) the citations to OT texts we know Mark used, yet didn't explicitly identify, as in the case of Daniel mentioned in n. 31 (p. 314).

54. For the now-fragmentary *Hazon Gabriel* (or Revelation of Gabriel) as a pos-sible foundational scripture for the Christians, see Israel Knohl, "'By Three Days, Live':

Messiahs, Resurrection, and Ascent to Heaven in Hazon Gabriel," *Journal of Religion* 88, no. 2 (April 2008): 147–58. It is likewise a known fact that many ancient manuscripts of the OT had variant readings unknown to us. The Dead Sea Scrolls revealed so many examples of new variants in the texts, peshers, and targums recovered there that we must conclude the quantity of still-lost variants is vast beyond reckoning, as that collection represents just a single library, and that of relatively small size.

55. For a discussion of this amusing literary fiasco, see Marcus Borg, "The Historical Study of Jesus and Christian Origins," in *Jesus at 2000*, ed. Marcus Borg (Boulder, CO: Westview, 1997), pp. 121–48 (this specific point on pp. 135–36).

56. Meier, *Marginal Jew*, vol. 1, pp. 168–69.

57. For more on the literary role of inventing a baptismal link between Jesus and John, see Thompson, *Messiah Myth*, pp. 33–37, 46–47; and Price, *Incredible Shrinking Son of Man*, pp. 101–30.

58. John Gager, "The Gospels and Jesus: Some Doubts about Method," *Journal of Religion* 54, no. 3 (July 1974):262–63, citing as an example the theory proposed in Morton Scott Enslin, *Christian Beginnings* (New York: Harper & Row, 1956), p. 156.

59. Mark 1:11; see Ehrman, *Orthodox Corruption*, pp. 47–118, for scribal attempts to conceal this in the manuscripts of Luke; that Mark probably underwent the same erasure is argued by the fact that Mark is clearly quoting Psalm 2:7, which contains the words now curiously erased from Mark, yet those missing words were fundamental to Christian tradition (Acts 13:33; Hebrews 1:5, 5:5) and obviously fit Mark's intentions. Meanwhile, Q cannot be invoked to rescue the baptism's historicity either: see Goodacre, "Criticizing the Criterion of Multiple Attestation."

60. Meier, *Marginal Jew*, vol. 1, p. 169.

61. See also *NIF*, pp. 369–72.

62. On *all* purported statements of Jesus predicting the end being possibly fabricated: Gager, "Gospels and Jesus," pp. 263–64. For the alternative view: John Loftus, "At Best Jesus Was a Failed Apocalyptic Prophet," *TCD*, pp. 316–46.

63. See Gager, "Gospels and Jesus," pp. 264–66. But see also Mogens Müller, *The Expression 'Son of Man' and the Development of Christology: A History of Interpretation* (Oakville, CT: Equinox, 2008) and Larry Hurtado and Paul Owen, eds., *Who Is This Son of Man? The Latest Scholarship on a Puzzling Expression of the Historical Jesus* (New York: T & T Clark, 2011). See also the additional sources (and yet another theory) provided in P. M. Casey, "Son of Man," in *The Historical Jesus in Recent Research*, ed. James Dunn and Scot McKnight (Winona Lake, IN: Eisenbrauns, 2005), pp. 315–24; but the analysis of Leslie Walck, *The Son of Man in the Parables of Enoch and in Matthew* (London: T & T Clark, 2011) is also now essential reading on this subject.

64. Meier, "Circle of the Twelve," p. 658.

65. Theissen and Winter, *Quest*, p. 175.

66. I am increasingly convinced there was no Q in the traditional sense, but the designation still conceptually defines *some* source, even if it turns out to be Matthew or some lost Gospel. I'll revisit this question in my next volume; but for now, I'll follow the widest consensus, which favors a Q tradition, although that consensus has been seriously challenged: see Mark Goodacre in *The Case against Q: Studies in Markan Priority and the Synoptic Problem* (Harrisburg, PA: Trinity, 2002); and his accompanying website at http://www.markgoodacre.org/Q.

67. Luke says Judas was replaced, though forty-seven days later (Acts 1:1–12: forty days plus a week = forty-seven days), which was also a full week *after* Luke says these appearances of Jesus had ended and Jesus had finally departed into the sky (Acts 1:15–26); this simply contradicts Paul's assertion that the very first people Jesus appeared to were Peter "and then the twelve" just *three* days later, only *then* followed by several more appearances to all the other brethren and apostles (last of all Paul: see 1 Corinthians 15:5–8). Someone is lying. Even if we conclude someone has doctored Paul's letter (see Robert Price, "Apocryphal Apparitions: 1 Corinthians 15:3–11 as a Post-Pauline Interpolation," *TET*, pp. 69–104), *they* would then be contradicting Luke. Luke's story is packed with implausibilities as it is. And there is ample reason to distrust Acts in general: see Richard Pervo, *The Mystery of Acts: Unraveling Its Story* (Santa Rosa, CA: Polebridge, 2008) and *Acts: A Commentary* (Minneapolis: Fortress Press, 2009). Possibly neither version is true, but the one that appears to come from Paul is the more credible (see TCD_w pp. 300–301 and 305–306), and at any rate must predate the Judas story, unless no contradiction was perceived by its author, which would negate an EC argument from the start. This is just one of countless examples of how so much of the evidence in Jesus studies is hopelessly problematic.

68. Fairly thoroughly established by the analysis of Haim Cohn, *The Trial and Death of Jesus* (New York: Harper & Row, 1971). It can be added that publicly crucifying a beloved demagogue on (or the day before) a high holiday (Mark 11:1–11; Matthew 21:1–11; Luke 19:29–44; John 12:12–19) would be so politically suicidal (virtually guaranteeing riots and violence in the city) that it's simply unbelievable. The prior probability that the Jewish elite would be that stupid is vanishingly small (a fact fully admitted by Mark, cf. 14:1–2, who nevertheless has them stupidly contradict themselves in the very next chapter, which is a sign of bad fiction more than honest history). It's *vastly* more likely that they would simply have jailed him, incommunicado, until the holiday passed, and conducted a trial then, when the vast throngs of visitors in Jerusalem had thinned and the passions of the holiday had passed.

69. Modern translations render his name familiarly as "Judas" when in fact his actual name in all NT documents is Judah (*Ioudas*). The word for Jew (*Ioudaios*) is the adjective of Judah, meaning "People of Judah" (hence "People of Judas"); likewise the word for the Holy Land, Judea (*Ioudaia*), means "Land of Judah" (hence "Land of Judas"). The latter

is of particular note considering the legend that land was bought with Judas's money and consecrated with his blood (if Matthew 27:3–10 and Acts 1:18–20 derive from a common story). In fact, according to Matthew, the land thus bought was declared to belong to "foreigners," in fact *dead* foreigners, and thus no longer Jewish, nor the land of the living. The symbolism is just too apposite to be anything but mythical (see following notes).

70. Compare Zechariah 11 (esp. in the LXX) and Matthew 27:3–10 (with possible allusions as well to Jeremiah 18:1–11 and 32:6–26). See Randel Helms, *Gospel Fictions*, pp. 112–17 and the relevant section of John Nolland, *The Gospel of Matthew: A Commentary on the Greek Text* (Grand Rapids, MI: William B. Eerdmans, 2005). Note that the thirty shekels Judas is paid in Matthew's version is exactly the legal value of a slave (Exodus 21:32), in fact a *dead* slave (thus it is what God's law declares you shall receive *in place of* a living servant). Credit for some of these observations is owed to Evan Fales, *Reading Sacred Texts: An Anthropological Approach to Matthew* (forthcoming). Even though I don't always agree with Fales, he has an astute eye for mytho-symbolic parallels.

71. The Zechariah tale has him giving the money "to the potter in God's temple" (Zechariah 11:13), so Matthew has Judas cast the money into God's temple (Matthew 27:5) and then the priests give it to the potter (for his field: 27:7); Acts 1:18 has Judas just buy a field (no details given). Matthew might also be alluding to Jeremiah 32 (see also Jeremiah 18–19), where Jeremiah is to buy a field and put the deed for it in a pot (Jeremiah 32:14, thus connecting a potter and a field), and that plot of land is saved while the sinners of the city are forsaken (Jeremiah 32:25); the Judas story reverses this: the buyers of the plot are to be forsaken (they are now foreigners who will inherit the grave—literally: Matthew 27:7), and the sinners of the city will be saved (in Jesus). That Matthew emphasizes how the Jewish elite in the end break their covenant with God (by their violation of the Sabbath) thus dovetails with his version of the Judas tale: cf. *TET,*, p. 362. Evan Fales also suggests that the priests in Matthew's account have in effect given away a parcel of the holy land (to foreigners—since burial establishes inheritance) in violation of God's covenant, and canceled their share of the atonement sacrifice (by taking back the money they paid for it). The Jewish elite are thus portrayed as total sell-outs who completely abdicate their participation in God's covenant. All of which would explain why Matthew even bothers to tell this story (as otherwise, what use is our knowing any of it?).

72. Dennis MacDonald also finds some literary parallels between the betrayer of Jesus and the betrayer of Odysseus: MacDonald, *Homeric Epics*, pp. 38–40 (cf. pp. 33–43). One can also see parallels in the betrayal of Joseph by *his* brother Judas in Genesis 37:18–36 (Joseph's cloak is taken and he is cast into a grave, sold to his enemies, and declared dead; Jesus' cloak is taken and he is cast into a grave, sold to his enemies, and declared dead; notably Joseph is betrayed by being sold into slavery, and in Matthew 27:9 Jesus is sold for the price of a slave); but also in the reversal of Israel's device of betraying his way *into* God's inheritance with a kiss (Genesis 27), while Judas betrays his way *out* of God's inheritance with a kiss—indeed,

in the OT the one kissed is Isaac, the sacrificed firstborn son *for whom* an animal is substituted (Genesis 22), while in the NT the one kissed is Jesus, the sacrificed firstborn son *substituted for* that same animal. We might not know if these allusions were intended, but they cannot simply be dismissed. They seem far too apposite to be mere coincidence.

73. For example, see the analysis of William John Lyons, "A Prophet Is Rejected in His Home Town (Mark 6.4 and Parallels): A Study in the Methodology (In) Consistency of the Jesus Seminar," *Journal for the Study of the Historical Jesus* 6, no. 1 (March 2008): 59–84.

74. The entire pericope thus looks artificial in Mark 3:21–30; John turns this single pericope into a running gag: John 7:20, 8:48, 8:52, 10:20 (accusations similarly leveled at John the Baptist, or so we're told: Matthew 11:18, Luke 7:33). That Christians were worried about being accused of madness is evinced in 1 Corinthians 14:23 and Acts 2:12–15, 26:24–25. It bears repeating that in Mark and John, Jesus is not said to *be* insane, but as being unjustly accused of it, which is not embarrassing to the self-righteous, who perceive themselves as facing such unjust accusations routinely. Further analysis of this pericope's literary function is provided in Lyons, "A Prophet Is Rejected in His Home Town."

75. See *NIF*, pp. 54–56.

76. I thoroughly discuss this fact in *NIF*, pp. 297–321 (where I also present several literary reasons for Mark inventing not only women in this story, but those particular women).

77. MacDonald, *Homeric Epics*, pp. 20–23.

78. Paul Danove, *The End of Mark's Story: A Methodological Study* (Leiden, Netherlands: E. J. Brill, 1993).

79. The closest thing we have to a statement of method in the Gospels appears in Luke 1:1, which is sometimes touted as indicating an effort at verification, but which when properly translated and understood actually entails the *absence* of any valid effort at verification: see *NIF*, pp. 178–82.

80. Note that $N(P)$ could be much larger than the number of such stories that actually survive for us to know of them now, because $N(\text{survive}) = N(P) \times n$, where $n =$ the percentage of stories that happened to survive for reasons unrelated to whether the story was fabricated or embarrassing. Since n does not discriminate between T and ~T it won't affect the ratio we're looking for—whatever the ratio of surviving true stories is to surviving false ones, it will be the same as the ratio of preserved true stories to preserved false ones (at least to a very high probability if $N(\text{survive})$ is sufficiently large, and even if it's small, any effect on that ratio will have been unpredictably random, and thus the percentage remains applicable, being what it is "so far as we know"). See chapter 6 (p. 214) for a discussion of coefficients of contingency (like n is here).

81. In other words, as many times as $N(\text{~T.M}) / [N(\text{~T.M}) + N(\text{~T.~M})]$ is greater than $N(\text{T.M}) / [N(\text{T.M}) + N(\text{T.~M})]$. Since $N(\text{~T.M}) = N(\text{~T})$ and therefore $N(\text{~T.~M}) = 0$, it follows that $N(\text{~T.M}) / [N(\text{~T.M}) + N(\text{~T.~M})] = 1$; and since $N(\text{T.M}) / [N(\text{T.M}) +$

N(T.~M)] = q, "as many times as N(~T.M) / [N(~T.M) + N(~T.~M)] is greater than N(T.M) / [N(T.M) + N(T.~M)]" becomes "as many times as 1 is greater than q."

82. Quoted in Porter, *Criteria for Authenticity*, p. 109.

83. Marcus Borg, "The Historical Study of Jesus and Christian Origins," *Jesus at 2000*, pp. 121–48 (quotes from pp. 145–46).

84. This criterion is discussed and criticized in Tuckett, "Sources and Methods," pp. 133–34; and Porter, *Criteria for Authenticity*, pp. 79–82; as well as in many of the critical references cited earlier.

85. Tuckett, "Sources and Methods," p. 134.

86. Anthony Le Donne, *The Historiographical Jesus: Memory, Typology, and the Son of David* (Waco, TX: Baylor University Press, 2009), p. 90.

87. See David Sim, *The Gospel of Matthew and Christian Judaism: The History and Social Setting of the Matthean Community* (Edinburgh, UK: T & T Clark, 1998); with David Sim, "Matthew's Use of Mark: Did Matthew Intend to Supplement or to Replace His Primary Source?" *New Testament Studies* 57, no. 2 (April 2011): 176–92; and "Matthew, Paul and the Origin and Nature of the Gentile Mission: The Great Commission in Matthew 28:16–20 as an Anti-Pauline Tradition," *Hervormde Teologiese Studies* 64, no. 1 (2008): 377–92.

88. See evidence and references in Margaret Williams, "VII.2. Pagans Sympathetic to Judaism" and "VII.3. Pagan Converts to Judaism" in *The Jews among the Greeks and Romans: A Diasporan Sourcebook* (Baltimore: Johns Hopkins University Press, 1998), pp. 163–79.

89. Tuckett, "Sources and Methods," pp. 134–35.

90. Ibid., p. 134. Many problems with this criterion are discussed in Goodacre, "Criticizing the Criterion of Multiple Attestation"; Porter, *Criteria for Authenticity*, pp. 82–89, 117–19; Eric Eve, "Meier, Miracle, and Multiple Attestation," *Journal for the Study of the Historical Jesus* 3.1 (2005): 23–45; and in Carrier, "Bayes' Theorem for Beginners," pp. 92–93.

91. See Herman Waetjen, *The Gospel of the Beloved Disciple: A Work in Two Editions* (New York: T & T Clark, 2005); C. K. Barrett, *The Gospel According to St. John*, 2nd ed. (Philadelphia: Westminster Press, 1978), pp. 15–26; C. H. Dodd, *Historical Tradition in the Fourth Gospel* (Cambridge: Cambridge University Press, 1963); also: *TET$_s$*, pp. 155–56, 191–93; Robert Price, *The Pre-Nicene New Testament* (Salt Lake City: Signature Books, 2006), pp. 665–718; and Andrew Gregory, "The Third Gospel? The Relationship of John and Luke Reconsidered," in *Challenging Perspectives on the Gospel of John*, ed. John Lierman (Tübingen, Germany: Mohr Siebeck, 2006), pp. 109–34, although arguing the reverse thesis, nevertheless summarizes the scholarship arguing the authors of John knew the Gospel of Luke; arguing a middle thesis (of shared sources), though still again cataloging evidence of dependence, is Raymond Brown and Francis Moloney, *An Introduction*

to the Gospel of John (Minneapolis: Doubleday, 2003) and Raymond Brown, *The Gospel According to John* (Minneapolis: Doubleday, 1966–1970).

92. Most persuasively argued by Goodacre, *Case against Q*, who also makes many direct and valuable points about fallacious methodology in the field.

93. On the extrabiblical evidence for Jesus and its paucity and problems, see Robert Van Voorst, *Jesus Outside the New Testament: An Introduction to the Ancient Evidence* (Grand Rapids, MI: William B. Eerdmans, 2000) and Gerd Theissen and Annette Merz, *The Historical Jesus: A Comprehensive Guide*, trans. John Bowden (Minneapolis: Fortress press, 1996).

94. Tuckett, "Sources and Methods," p. 134.

95. Borg, "Historical Study," pp. 144–45.

96. Most importantly argued in Michael Grant, *Greek and Roman Historians: Information and Misinformation* (New York: Routledge, 1995).

97. Strauss, *Four Portraits, One Jesus*, p. 262. A better formulation of this criterion is found in H. W. Shin, *Textual Criticism and the Synoptic Problem in Historical Jesus Research: The Search for Valid Criteria* (Dudley, MA: Peeters, 2004), pp. 167–90, yet is there more clearly just an unwitting restatement of BT (in fact almost wittingly, e.g., pp. 210–18: although on p. 218 he gives an invalid mathematical model, violating basic principles of probability theory, correcting him produces BT).

98. Tuckett, "Sources and Methods," p. 135; see also Marcus Borg, *Jesus: Uncovering the Life, Teachings, and Relevance of a Religious Revolutionary* (New York: HarperSanFrancisco, 2006), pp. 72–73. Notably, this criterion can easily conflict with the Criterion of Dissimilarity, a methodological problem in itself (as I noted earlier).

99. For instance, the increasingly accepted theory that Luke used the writings of Josephus to precisely that purpose: Richard Pervo, *Dating Acts: Between the Evangelists and the Apologists* (Santa Rosa, CA: Polebridge, 2006); Steve Mason, "Josephus and Luke-Acts," *Josephus and the New Testament* (Peabody, MA: Hendrickson, 1992), pp. 185–229; Gregory Sterling, *Historiography and Self-Definition: Josephos, Luke-Acts and Apologetic Historiography* (Leiden, Netherlands: Brill, 1992); Heinz Schreckenberg, "Flavius Josephus und die lukanischen Schriften," in *Wort in der Zeit: Neutestamentliche Studien*, Karl Rengstorf and Wilfrid Haubeck, eds. (Leiden, Netherlands: Brill, 1980), 179–209; Max Krenkel, *Josephus und Lucas: Der Schriftstellerische Einfluss des Jüdischen Geschichtschreibers auf den Christlichen* (Leipzig, Germany: H. Haessel, 1894).

100. Tuckett, "Sources and Methods," p. 136.

101. Porter, *Criteria for Authenticity*, pp. 116–22, discusses problems with this criterion and the previous two, which he treats as special cases of this one.

102. Borg, *Jesus at 2000*, pp. 73–75.

103. Borg, "Historical Study of Jesus," p. 145. On the late dating of the final edition of John (the one we have), see scholarship cited in note 91 (p. 320),

104. For example: Eric Kandel, *In Search of Memory: The Emergence of a New Science of Mind* (New York: W. W. Norton, 2006); C. J. Brainerd and V. F. Reyna, *The Science of False Memory* (New York: Oxford University Press, 2005); Alan Baddeley, *Your Memory: A User's Guide*, new illustrated ed. (Buffalo, NY: Firefly, 2004); Daniel Schacter, *The Seven Sins of Memory: How the Mind Forgets and Remembers* (Boston: Houghton Mifflin, 2001); Daniel Schacter and Joseph Coyle, eds., *Memory Distortion: How Minds, Brains, and Societies Reconstruct the Past* (Cambridge, MA: Harvard University Press, 1995); Elizabeth Loftus and James Doyle, *Eyewitness Testimony: Civil and Criminal*, 3rd ed. (Charlottesville, VA: Lexis Law, 1997); and Gary Wells and Elizabeth Loftus, eds., *Eyewitness Testimony: Psychological Perspectives* (New York: Cambridge University Press, 1984). Experts on the reliability of oral traditions have always concurred, noting specifically that oral traditions in which we no longer have access to a living tradent are particularly suspect (such as we have in the Gospels, e.g., we cannot interview their authors or ascertain their chain of transmission): Jan Vansina, *Oral Tradition: A Study in Historical Methodology* (London: Routledge, 1965); Paul Richard Thompson, *The Voice of the Past: Oral History* (New York: Oxford University Press, 1978); David Henige, *Oral Historiography* (New York: Longman, 1982); Rosalind Thomas, *Oral Tradition and Written Record in Classical Athens* (New York: Cambridge University Press, 1989). On applying these findings to the sources for Jesus, see Dale Allison Jr., *Constructing Jesus: Memory, Imagination, and History* (Grand Rapids, MI: Baker Academic, 2010).

105. Tuckett, "Sources and Methods," p. 136. Problems with this criterion are discussed by Porter, *Criteria for Authenticity*, pp. 110–13.

106. Borg, "Historical Study of Jesus," p. 145. This is essentially the converse of the Criterion of Embarrassment (in which claims *against* the fabricatory trend are assumed to be *more* true—which, as we saw, is too simplistic).

107. For quote and critique: Porter, *Criteria for Authenticity*, pp. 77–79.

108. Phenomena now well understood scientifically. On the science of memory, see note 104 (p. 322).

109. See discussion and references in *NIF*, pp. 161–87, and *SGG*, pp. 246–47.

110. See Porter, *Criteria for Authenticity*, pp. 181–209. This and the following three criteria are the very ones refuted by Avalos and Bird, and then conceded even by Porter as not demonstrating historicity after all (see nn. 5 and 6 in chap. 1, p. 293).

111. A demonstrated phenomenon in the Gospels: Mark Goodacre, *Case against Q*, pp. 40–43.

112. Porter, *Criteria for Authenticity*, pp. 127–80 (quoting pp. 143–44). See note 110.

113. Ibid., pp. 89–100. Porter is more critical of this criterion than his own, even though many of the same problems attend.

114. Tuckett, "Sources and Methods," p. 136.

115. Porter, *Criteria for Authenticity*, p. 79 (quoting E. P. Sanders).

116. Porter, *Criteria for Authenticity*, pp. 210–37 (quoting p. 217). See note 110 on the Criterion of Textual Variance. Discourse features analysis can be a valid technique, but requires far better source materials than we have for Jesus, cf., e.g., Robert Eagleson, "Forensic Analysis of Personal Written Texts: A Case Study" and Wilfrid Smith, "Computers, Statistics and Disputed Authorship," in *Language and the Law*, ed. John Gibbons (New York: Longman, 1994), pp. 362–73 and 374–413. Similarly: Erica Klarreich, "Bookish Math: Statistical Tests Are Unraveling Knotty Literary Mysteries," *Science News* 164, no. 25–26 (December 20 and 27, 2003): 392–93; Donald Foster, *Author Unknown: Tales of a Literary Detective* (New York: H. Holt, 2000); Ian Marriott, "The Authorship of the *Historia Augusta*: Two Computer Studies," *Journal of Roman Studies* 69 (1979): 65–77.

117. James Dunn, "The Characteristic Jesus," *A New Perspective on Jesus: What the Quest for the Historical Jesus Missed* (Grand Rapids, MI: Baker Academic, 2005), pp. 69–78 (quoting p. 69).

118. On this and other ancient educational practices: David Gowler, "The Chreia," in Levine, Allison, and Crossan, *Historical Jesus in Context*, pp. 132–48; Tim Whitmarsh, *Greek Literature and the Roman Empire: The Politics of Imitation* (New York: Oxford University Press, 2001); Raffaella Cribiore, *Gymnastics of the Mind: Greek Education in Hellenistic and Roman Egypt* (Princeton, NJ: Princeton University Press, 2001); MacDonald, *The Homeric Epics and the Gospel of Mark*, pp. 4–6 and *Christianizing Homer: The Odyssey, Plato, and the Acts of Andrew* (New York: Oxford University Press, 1994); and extensively in Thomas Brodie's doctoral dissertation, "Luke the Literary Interpreter: Luke-Acts as a Systematic Rewriting and Updating of the Elijah-Elisha Narrative in 1 and 2 Kings" (Pontifical University of St. Thomas Aquinas, 1981), pp. 5–93.

119. For example: Robert Stein, "Criteria for the Gospels' Authenticity," *Contending with Christianity's Critics: Answering New Atheists and Other Objectors*, eds. Paul Copan and William Lane Craig (Nashville, TN: B & H Academic, 2009), pp. 88–103 (cf. p. 98).

120. Employed throughout Anthony Le Donne, *The Historiographical Jesus* and *Historical Jesus: What Can We Know and How Can We Know It?* (Grand Rapids, MI: William B. Eerdmans 2011).

121. Le Donne, *Historical Jesus*, p. 37.

122. Ibid., p. 78.

123. Ibid., pp. 91–92; his dependence on the same invalid criteria to distinguish memories from inventions is explicitly stated, for example, in *Historiographical Jesus*, pp. 82 and 86–91.

124. Ibid., pp. 126–31.

125. See 1 Corinthians 3:15–17 and 6:19 (even the Pseudo-Pauline Ephesians 2:20–22).

126. See *TET$_s$*, pp. 139–47 and 156–57 (with 219, nn. 263 and 264). Curiously Luke represents the saying (that Jesus would destroy the temple) as originating with Stephen, not

Jesus (Acts 6:14), but in so doing completely erases its metaphor, dropping the elements of "in three days" and "building with hands/without hands," even though Luke certainly knew they belonged, having used Mark as his source—yet he deliberately deleted this material from the trial and crucifixion. He might thus be deliberately erasing the metaphor altogether (as he would be inclined to do, having a different theology of resurrection than Paul's, per the analysis in *TET₃*), and in Acts only alluding instead to Luke 21, which redacts Mark 13, where Jesus literally, not metaphorically, refers to the temple being destroyed—but not to his doing it, nor to rebuilding it, much less in three days, and without any mention of these temples being made with or without hands; whereas all those details, obvious Pauline markers, are carefully included by Mark in verse 14:58, which makes this far more likely a literary invention of Mark (or his Pauline sources). Notably the saying is only actually placed within Jesus' ministry in John, the last and least reliable Gospel to be written. Mark never places it in his ministry, but instead has it reported secondhand as having been there (twice, in both cases illustrating the ironic failure of his enemies to "get" the secret Pauline meaning of the saying, in accord with Mark 4:11–12, 33–34; Matthew then fumbles the irony by having his enemies "get" it after all, Matthew 27:61–65, although that may have been intentional, because it generated a new irony: *TET₁*, p. 362).

127. Le Donne, *Historical Jesus*, p. 80.

128. Quotation from ibid., p. 43.

129. Strauss, *Four Portraits, One Jesus*, pp. 360–65 (quoting pp. 361 and 362).

130. Dennis MacDonald, "Imitations of Greek Epic in the Gospels," in Levine, Allison, and Crossan, *Historical Jesus in Context*, pp. 372–84 (quoting p. 374). On these criteria and their application, as well as more on the teaching of mimesis in ancient schools, see Dennis MacDonald, *Homeric Epics; Does the New Testament Imitate Homer? Four Cases from the Acts of the Apostles* (New Haven, CT: Yale University Press, 2003); "The Shipwrecks of Odysseus and Paul," *New Testament Studies* 45 (1999): 88–107; and "Secrecy and Recognitions in the *Odyssey* and Mark: Where Wrede Went Wrong," *Ancient Fiction and Early Christian Narrative*, ed. Ronald Hock, J. Bradley Chance, and Judith Perkins (Atlanta, GA: Scholars Press, 1998), pp. 139–54. Also proving the same results are the published works of Thomas Brodie, e.g., *The Birthing of the New Testament: The Intertextual Development of the New Testament Writings* (Sheffield, UK: Sheffield Phoenix Press, 2004), and Randel Helms, e.g., *Gospel Fictions* (Amherst, NY: Prometheus Books, 1988); see also Dennis MacDonald, ed., *Mimesis and Intertextuality in Antiquity and Christianity* (Harrisburg, PA: Trinity Press International, 2001). I will survey some of their examples and more from other published scholars in my next volume.

131. Thomas Brodie, *Proto-Luke: The Oldest Gospel Account: A Christ-Centered Synthesis of Old Testament History Modeled Especially on the Elijah-Elisha Narrative: Introduction, Text, and Old Testament Model* (Limerick, Ireland: Dominican Biblical Institute, 2006), p. 3.

132. *TET₁*, pp. 349–68.

133. Thomas Mathews, *The Clash of the Gods: A Reinterpretation of Christian Art* (Princeton, NJ: Princeton University Press, 1993), pp. 66, 72, 77–84 (with figs. 55, 59); Robin Margaret Jensen, *Understanding Early Christian Art* (New York: Routledge, 2000), pp. 174–78 (with figs. 21, 24).

134. Against biblical fundamentalists who protest against this obvious conclusion, see *TET₁*, pp. 358–59, and my more complete response in my online FAQ for that chapter, at http://www.richardcarrier.info/TheftFAQ.html#harmonization.

135. On generic prediction, see chapters 3 (p. 77) and 6 (p. 214). On Matthew's literary aims in converting Jesus' tomb into Daniel's den, see *TET₁*, pp. 358–64.

CHAPTER 6. THE HARD STUFF

1. David Hackett Fischer, *Historians' Fallacies: Toward a Logic of Historical Thought* (New York: Harper & Row, 1970), p. 264.

2. And if you're more mathematically inclined, sophisticated techniques for pooling disagreeing probability estimates from informed experts have also been developed and shown to actually increase accuracy, and such techniques could be employed even when disagreement persists, in order to generate a result that will likely be even more correct than any single expert opinion. See Thomas Wallsten and Adele Diederich, "Understanding Pooled Subjective Probability Estimates," *Mathematical Social Sciences* 41, no. 1 (January 2001): 1–18. However, this only works well when all the experts in the pool are more or less equally informed.

3. The reasoning (and scientific and cultural facts) underlying this point are thoroughly discussed in *TCD*.

4. *SGG*, pp. 11–14 ("mystically" experiencing the Tao); *TET₃*, pp. 184–88 ("physically" combating a demon), and for the science pertaining to the latter see Bruce Bower, "Night of the Crusher: The Waking Nightmare of Sleep Paralysis Propels People into a Spirit World," *Science News* 168, no. 2 (July 9, 2005): 27–29.

5. As suggested throughout this book, historians should classify as *e* the specific evidence that requires explanation (i.e., that *h* purports to explain), and classify everything else as *b* (i.e., everything all historians know or should be able to know). As long as logical consistency is maintained, the evidence can be divided in any way between *b* and *e*, and BT will always generate the same conclusion. I'll demonstrate this further below.

6. I also say more about improving probability estimates in my tutorial at http://www.richardcarrier.info/CarrierDec08.pdf (cf. pp. 36–38).

7. This same problem has also vexed a number of philosophers of history, cf. C. Behan McCullagh, *Justifying Historical Descriptions* (New York: Cambridge University Press,

1984), pp. 33–38 (McCullagh resolves the problem differently, but his solution can be reduced to the solution entailed by BT, which I shall here describe).

8. The priors don't help either (for a variety of reasons they would heavily favor the Big Bang over God in this case), but here I'm concerned only with the consequents.

9. In fact scientists even decide between different versions of the Big Bang Theory by looking for different patterns of radiation in the star field (where the *exact* pattern of, for example, the microwave background radiation is not what different theories predict, but only differences in *generic features* of that background).

10. Although scientists already get the point: see Sérgio B. Volchan, "Probability as Typicality," *Studies in History and Philosophy of Science Part B: Studies in History and Philosophy of Modern Physics* 38, no. 4 (December 2007): 801–814. Volchan essentially demonstrates mathematically the entire point I make here, only using statistical mechanics as an extended example.

11. Indeed, information in *b* might even entail that an observation of Dr. Smith is more likely than Dr. Jones due to such facts as that Dr. Jones is on vacation, or missing and presumed dead. And yet if Dr. Jones makes the observation anyway, it would be perverse to allow this to render *h* less probable merely because Dr. Jones's being the one to make the observation is less probable—because that would have nothing to do with *h* being true or false.

12. McCullagh, *Justifying Historical Descriptions*, pp. 33–38.

13. This is thoroughly discussed (using BT) in my own TEC_d, which relies on the even more thorough mathematical demonstrations of Michael Ikeda and Bill Jefferys, "The Anthropic Principle Does Not Support Supernaturalism," bayesrules.net/anthropic .html, an earlier version of which appeared in Michael Martin and Ricki Monnier, eds., *The Improbability of God* (Amherst, NY: Prometheus Books, 2006), pp. 150–66; and Elliott Sober, "The Design Argument," philosophy.wisc.edu/sober/design argument 11 2004. pdf, an earlier version of which is in the 2004 edition of Charles Taliaferro, Paul Draper, and Philip Quinn, eds., *A Companion to Philosophy of Religion* (Cambridge, MA: Wiley-Blackwell), pp. 117–48.

14. See my essay (and the references therein): Richard Carrier, "Statistics & Biogenesis," May 1, 2009, at http://richardcarrier.blogspot.com/2009/05/statistics-biogenesis _01.html.

15. As I prove specifically in TEC_d, pp. 289–92.

16. According to Laplace's Rule of Succession: $(s + 1) / (n + 2) = (67 + 1) / (67 + 2) = 68 / 69 = 0.98550$ (rounding off at the fifth decimal place), which is roughly 98.6 percent. We could be wrong about that. Due to random fluctuations in the data, the actual probability could be different than Laplace's Rule predicts, but with such a large sample that becomes increasingly unlikely (see discussion of hypothetical vs. actual frequencies later in this chapter). Moreover, an *a fortiori* estimate of 'at least 95%' substituted for the exact

result of 98.6% would be accurate to an even higher probability. Likewise the other case, of colonies without special patronage, if we had 100 prior cases and 20 had libraries, then $(s + 1) / (n + 2) = (20 + 1) / (100 + 2) = 21 / 102 = 0.20588$, or roughly 21%, not strictly the 20% you might expect $(20/100 = 20\%)$, but such a small difference is rarely significant, and diminishes as the population size increases, such that for a hypothetical set containing infinite elements (i.e., every logically possible member) the probability is simply 20%.

17. This problem is discussed by John Earman, *Bayes or Bust? A Critical Examination of Bayesian Confirmation Theory* (Cambridge, MA: MIT Press, 1992), pp. 139–41.

18. I realize the whole of the above simply colloquializes a lot of what could be stated in exact mathematical terms (which, especially when given the original data, could show that the prior probability in this example should indeed be heavily weighted in favor of P(NP)), but the aim of this book is to convey the logic that historians need to use, without pressing upon them excessively complicated mathematics that they won't really require in order to reach the same conclusions with enough approximation to suit their needs. Nevertheless, someday historians might start employing detailed mathematical arguments when such is necessary and useful to establishing their conclusions, and I encourage that— as long as everything is intelligibly explained so their peers can verify the steps in their argument.

19. Though mathematically this is identical to using 0.5 as the prior (or anything else we determine from any background evidence that we reserve for determining the prior) and then multiplying all consequents for all cases—and sometimes that's just easier, e.g., we can run a series $BT_1 \rightarrow BT_2 \rightarrow BT_3$ or just run one $BT_{1,2,3}$ by substituting for P(e|h.b) the sequence $P(e_1|h.b) \times P(e_2|h.b) \times P(e_3|h.b)$, and substituting for P(e|~h.b) the sequence $P(e_1|\sim h.b) \times P(e_2|\sim h.b) \times P(e_3|\sim h.b)$. For example, if P(h|b) = 0.8, $P(e_1|h.b) = 0.8$, $P(e_2|h.b) = 0.7$, $P(e_3|h.b) = 0.6$, and $P(e_1|\sim h.b) = 0.5$, $P(e_2|\sim h.b) = 0.4$, and $P(e_3|\sim h.b) = 0.3$, then using the step-by-step method $BT_1 = 0.8 \times 0.8 / (0.8 \times 0.8) + (0.2 \times 0.5) = 0.8649$, and thus $BT_2 = 0.8649 \times 0.7 / (0.8649 \times 0.7) + (0.1351 \times 0.4) = 0.9181$, and thus $BT_3 = 0.9181 \times 0.6 / (0.9181 \times 0.6) + (0.0819 \times 0.3) = 0.9573$, which is the final posterior probability once all evidence is considered. Comparatively, using the all-at-once method, $BT_{1,2,3} = 0.8 \times (0.8 \times 0.7 \times 0.6) / (0.8 \times (0.8 \times 0.7 \times 0.6)) + (0.2 \times (0.5 \times 0.4 \times 0.3)) = 0.9573$. The same result. So whether you use one equation or a nested series of equations is only a matter of which approach is more convenient to the occasion; otherwise they produce identical results.

20. Because if $BT_1 \rightarrow BT_2 \rightarrow BT_3 = BT_{1,2,3}$ (and, as shown previously by example, it is), then the converse is true $(BT_{1,2,3} = BT_1 \rightarrow BT_2 \rightarrow BT_3)$; a case where there is one pair of consequents to calculate is mathematically identical to multiplying those consequents against a pair of consequents equal to 1 (representing all the other evidence that is just as entirely expected on h as on $\sim h$); and if BT_1 contains only the latter evidence (of consequent probability 1) and starts with a neutral prior, then its e will entail a posterior probability of 0.5

for BT_1, and if BT_2 contains all the other evidence, then it also has a prior of 0.5 (being the posterior probability of BT_1). We are positing that BT_2 has a P(SILENCE|h.b) = 0.2 and a P(SILENCE|~h.b) = 0.6; by the law of commutation $BT_1 \rightarrow BT_2 = BT_2 \rightarrow BT_1$ (i.e., it doesn't matter to the conclusion in what order you will examine the evidence); so if we started with the final probability of BT_2 that would become the prior probability in BT_1 and as they must, when combined, entail the same final posterior probability (the output of $BT_{1,2}$), this means the *consequents* of BT_2 entail a *prior probability*, as in this case they entail what becomes the prior probability in BT_1 (and as for this case, so for all others), ergo every pair of consequents entails some reference class from which a prior can be derived instead (if, e.g., we wanted to use the evidence in BT_2 to construct a reference class and derive a prior probability from it, rather than to construct a pair of consequent probabilities from that same evidence and derive our prior probability from another set of data).

21. The reason why *a fortiori* estimates avoid the problem is that they negate the effect of any inconsistency that might arise from them, i.e., if we recalibrated our estimates to produce strict consistency, our revised final probability will always *more* strongly support the conclusion, so we don't need to ensure strict consistency. In other words, the practice of arguing *a fortiori* ensures our inconsistencies will always be in the same direction, the one direction that won't affect our overall conclusion (e.g., if the conclusion is "P(h|e.b) is more than 0.8" and we used *a fortiori* estimates to arrive at that conclusion, no enforcement of consistency would alter that conclusion, e.g., doing so might revise our result to "P(h|e.b) is more than 0.9" but since 0.9 *is already* more than 0.8, the original conclusion remains correct). Of course, if we are inconsistent in assigning probabilities even *a fortiori*, then we'll always be wrong no matter what we do, but such inconsistencies usually won't require mathematical tests to detect.

22. Per the preceding example, $BT_1 \rightarrow BT_2 \rightarrow BT_3 = BT_3 \rightarrow BT_2 \rightarrow BT_1 = BT_3 \rightarrow BT_1 \rightarrow BT_2$ = etc. Thus one reference class in this series can always simply be diverted into developing a pair of consequent probabilities in that series instead, so it won't matter which class you start with to determine the prior. The final probability will, in the end, always be the same. Because, simply put, *b* and *e* always exhaust all knowledge available to you, *no matter how you divide the evidence between them.*

23. Indeed the number of persons claimed to have been thus raised in antiquity is well more than two dozen: see *NIF*, pp. 85–127. And those are just the ones we know about.

24. See my essay: Richard Carrier, "Our Mathematical Universe," at http://richard carrier.blogspot.com/2007/10/our-mathematical-universe.html, published October 5, 2007.

25. Since all those superfluous hypotheses have a P(e|h.b) of 0 (or near enough), it is possible to ignore them when using the odds form of Bayes's Theorem (p. 284), since then you only need to consider the ratio of the priors among the remaining hypotheses. And you can also ignore them when using any form of Bayes's Theorem by setting the highest consequent (among all contenders) as equal to 1 and setting all the remaining consequents in pro-

portion to that (in the same way we could for the consequent possibilities in the Trustworthy Neighbor example, on page 74). The priors of disregarded hypotheses are then folded into ~h.

26. This should warn you against ever assigning a prior of exactly 1 or 0, because doing so commits a textbook fallacy of begging the question (i.e., it produces a circular argument) since a 1 means *all* elements of a set match our h (and 0 means that *none* do), but one of the elements of that set is obviously any h we are testing, and we can't begin by *presuming* that our hypothesis is true (or false). There would be no point in continuing the analysis were that the case. More importantly, such certainty for us is *logically impossible* (at least for all substantive claims about history; for propositions that really do have a prior of 0 or 1, see chap. 2, p. 23), therefore it would be a violation of logic to assert that a prior probability was exactly 0 or 1. We can use (or presume) a 0 or a 1 in an actual working equation only with the understanding that *on a higher resolution analysis* that 0 or 1 would actually be a number *very near* (but not exactly) 0 or 1, and not actually in fact 0 or 1 (except in self-evident cases of logical necessity, where we don't need BT, because we already know a proposition cannot be true or cannot be false).

27. See Sérgio B. Volchan, "Probability as Typicality," *Studies in History and Philosophy of Science Part B: Studies in History and Philosophy of Modern Physics* 38, no. 4 (December 2007). In formal mathematical notation BT is actually represented by:

$$P(h_0|e.b) = \frac{P(h_0|b)\, P(e|h_0.b)}{\sum_n P(h_n|b)\, P(e|h_n.b)}$$

This represents the fact that: BT is P(h|e.b) = [P(h|b) × P(e|h.b)] / ([P(h|b) × P(e|h.b)] + [P(~h|b) × P(e|~h.b)]), which equals [P(h|b) × P(e|h.b)] / ([P(h_0|b) × P(e|h_0.b)] + [P(h_1|b) × P(e|h_1.b)] + [P(h_2|b) × P(e|h_2.b)] + [P(h_3|b) × P(e|h_3.b)] + . . . + [P(h_n|b) × P(e|h_n.b)]) for all possible hypotheses, which entails that [P(h|b) × P(e|h.b)] + [P(~h|b) × P(e|~h.b)] = [P(h_0|b) × P(e|h_0.b)] + [P(h_1|b) × P(e|h_1.b)] + [P(h_2|b) × P(e|h_2.b)] + [P(h_3|b) × P(e|h_3.b)] + . . . + [P(h_n|b) × P(e|h_n.b)], which entails that, when $h = h_0$, then [P(~h|b) × P(e|~h.b)] = [P(h_1|b) × P(e|h_1.b)] + [P(h_2|b) × P(e|h_2.b)] + [P(h_3|b) × P(e|h_3.b)] + . . . + [P(h_n|b) × P(e|h_n.b)], therefore removing any hypothesis (e.g., [P(h_1|b) × P(e|h_1.b)]) from ~h must necessarily reduce the result of [P(~h|b) × P(e|~h.b)] by exactly as much as would account for it if it were the one being tested instead (i.e., if it became h_0), therefore [P(~h|b) × P(e|~h.b)] must include all possible hypotheses besides h. The value of [P(~h|b) × P(e|~h.b)] for all those other hypotheses must then equal the value of [P(h_1|b) × P(e|h_1.b)] + [P(h_2|b) × P(e|h_2.b)] + [P(h_3|b) × P(e|h_3.b)] + . . . + [P(h_n|b) × P(e|h_n.b)]. And this is what entails that the sum of prior probabilities for all hypotheses must equal 1: P(h_0|b) + P(h_1|b) + P(h_2|b) + P(h_3|b) + . . . + P(h_n|b) = 1, because P(h|b) + P(~h|b) = 1, since h and ~h exhaust all logical possibilities, and the probability is 100% that the truth is one of all the possible things there are, and because $P(h|b) = P(h_0|b)$, and $P(\sim h|b) = P(h_1|b) + P(h_2|b) + P(h_3|b) + \ldots + P(h_n|b)$. Therefore only hypotheses whose removal from $\sim h$ would have no visible effect on our math (i.e., any hypothesis n for which $[P(h_n|b) \times P(e|h_n.b)] \approx 0$) can be safely ignored.

28. Although I selected this example precisely because it is *not* physically impossible to transmute lead into gold (and thus we're not facing any mere 'bias against the supernatural' here as discussed in chap. 4, p. 114). Rather, our b entails it's just not possible for anyone of ancient Galilee to have done this, not having had any of the technological infrastructure or scientific knowledge known to be necessary to accomplish it. Gold is far more efficiently transmuted from mercury or platinum, but can be transmuted from lead by simply knocking out a few protons. But the only way to do that is to employ a particle accelerator or nuclear reactor; and to produce enough gold to get rich on, one would have to apply these technologies on a remarkably vast scale (and even then the gold produced might be so radioactive as to kill its possessor). As we know of no other way to do it, and in fact know a great deal about what prevents it being done any other way, we are correct in assigning it a vanishingly small prior probability in the case of Matthias (until, of course, we discover some new physics or power).

29. More precisely, the prior in this case would be 99.7%, according to Laplace's Rule: $(998+1) / (1000 + 2) = 999 / 1002 = 0.997$ (rounded), but for a hypothetical set of infinite extension, it's 99.8%. Meanwhile the consequent probability in this analysis is simply 1 (for both h and $\sim h$), so the prior probability is also the posterior probability, and thus simply the probability. Of course the same problem can be analyzed the other way around, but it would produce the same results: if 99.8% amazing hands are fair, and "amazing hand" is defined as having a probability of 1 in 100,000 on a fair deal, then $P(e|\text{CHEAT}.b) = 1$ (assuming a cheater would want no other outcome), while $P(e|\text{FAIR}.b) = 0.00001$ (the 1 in 100,000 natural odds); but on that model, "0.998 amazing hands are fair" then entails $P(\text{CHEAT}|b) = 0.00000002$ (i.e., only one in fifty million of all hands dealt can then be cheats—meaning all hands whatever, whether amazing or not), so $P(\text{FAIR}|e.b)$ would still equal 99.8%.

30. An example of a "no one ever thought of that" factor is a recent study proving that coin tosses are not random but always slightly favor the side of the coin that was facing up before the toss: Erica Klarreich, "Toss Out the Toss-Up: Bias in Heads-or-Tails," *Science News* 165, no. 9 (February 28, 2004): 131.

31. For more on all this and its broader philosophical relevance, see my essay: Richard Carrier, "Our Mathematical Universe," at http://richardcarrier.blogspot.com/2007/10/our -mathematical-universe.html, published October 5, 2007.

32. On the reason we reject that 'aliens' hypothesis, see the concluding section of my essay: Richard Carrier, "Defining the Supernatural," January 18, 2007, at richardcarrier .blogspot.com/2007/01/defining-supernatural.html. That essay elaborates a more general argument later published in Richard Carrier, "On Defining Naturalism as a Worldview," *Free Inquiry* 30, no. 3 (April/May 2010): 50–51.

33. According to Laplace's Rule of Succession: $(s + 1) / (n + 2) = (4 + 1) / (4 + 2) = 5 / 7 = 71.4\%$.

34. Some of the underlying mathematics is discussed in William Faris's book review of *Probability Theory: The Logic of Science* (by E. T. Jaynes), in *Notices of the American Mathematical Society* 53, no. 1 (January 2006): 33–42.

35. And as Giulio D'Agostini demonstrates, no one really does, not even the most stalwart of frequentists: "Teaching Statistics in the Physics Curriculum: Unifying and Clarifying Role of Subjective Probability," *American Journal of Physics* 67, no. 12 (December 1999): 1261–62.

36. A funerary inscription attests an industrial mechanic who achieved rather considerable wealth under the Roman Empire in what is now modern Turkey: Tullia Ritti, Klaus Grewe, and Paul Kessener, "A Relief of a Water-Powered Stone Saw Mill on a Sarcophagus at Hierapolis and Its Implications," *Journal of Roman Archaeology* 20 (2007): 138–63.

37. In this case, of course, we might not be talking about the prior probability of someone getting rich by being an industrial mechanic, but the prior probability of this specific evidence (the funerary inscription) being produced by its occupant's success in industrial mechanics, rather than by this evidence (the inscription) being forged, or the inscribers lying or being mistaken, etc., although the one frequency does relate to the other (see n. 11 for chap. 3, p. 301, on the effect of the frequency of events on the believability of testimony).

38. See note 4 for chapter 2, p. 297, regarding the illogical terminological equation of these two ideas of probability with the phrases 'objective probability' and 'subjective probability' respectively (which convention I believe should be abandoned, for the reasons stated there).

39. Almost every other dispute between frequentists and Bayesians is quickly dispatched by Giulio D'Agostini, "Role and Meaning of Subjective Probability: Some Comments on Common Misconceptions," October 26, 2000, http://arxiv.org/abs/physics/0010064. D'Agostini demonstrates the frequentists are engaged in far more folly than the Bayesians. But he stops just short of the last step I take here, which finally folds everything that's correct about frequentism into Bayesianism, thus eliminating everything frequentists claim is incorrect about Bayesianism.

40. Faris's book review of *Probability Theory*, p. 36.

41. A confidence level of 100% is mathematically and logically impossible, as we never have access to 100% of all information, i.e., we're not omniscient, and as Gödel proved, no one can be, for it's logically necessary that there will always be things we won't know, even if we're God—like whether a belief that we're omniscient and infallible is true, which cannot be noncircularly asserted as known with 100% confidence, even by God.

42. Formally speaking: if $P(r) \to 1$, then $(1 - P(r)) \to 0$ and therefore $[(1 - P(r)) \times P(d')] \to 0$, because $\{(\to 0) \times P(d')\} \to 0$ (since anything times zero is zero); ergo the entire factor

"+ $[(1 - P(r)) \times P(d')]$" can effectively be ignored, leaving $P(h|b) = P(r) \times P(d)$, and if $P(r)$ →1, then $P(h|b) = P(r) \times P(d) = (→1) \times P(d)$, and $((→1) \times P(d))→P(d)$; ergo $P(h|b)→P(d)$. Which also means the fact that (obviously) we almost never know $P(d')$ is irrelevant—as its value can make no difference to this result: because $P(d')$ only differs from $P(d)$ to a probability of $1 - P(r)$, so when $P(r)$ is extremely high (approaching 1), $P(d')$ will almost certainly not differ significantly from $P(d)$, that is, it will do so only with extreme rarity

43. Of course, if we were *truly* omniscient and infallible, we wouldn't need to run a BT analysis, as we'd already know the answer. So assume that by 'omniscient' we only know general and not specific facts (like the hero of the ill-fated television series *John Doe*), e.g., we know how many people won the lottery, but not exactly who those people are.

44. The example being referred to is discussed in C. Behan McCullagh, *Justifying Historical Descriptions* (New York: Cambridge University Press, 1984), p. 22.

45. John Earman, *Bayes or Bust?: A Critical Examination of Bayesian Confirmation Theory* (Cambridge, MA: MIT Press, 1992), p. 114.

46. Which is essentially Earman's solution to the problem, cf. *Bayes or Bust?* p. 117 (cf. also 120ff. and 158).

47. Although it's worth noting his geocentric model was also very *predictively* successful, only failing after a very long time (with the notable exception of apparent diameters, which never conformed to the theory, past or future).

48. See discussion in chapter 3 (p. 79) of the folly of trying to "ignore" evidence (leaving it out of both *e* and *b*), which can never be logically valid.

INDEX

ABE. *See* Argument to the Best
Explanation
ad hocness, 101, 102
ad hoc theory reinforcement. *See* ad
hocness; gerrymandering (a theory)
AFE. *See* Argument from Evidence
affective fallacy, 188
a fortiori, method of, 57, 78, 84,
85–88, 93–95, 100, 111, 198, 209,
213, 218, 235, 238–39, 241–42,
248, 252, 261, 263, 264, 268–70,
275, 280, 283, 304n20, 326–27n16,
328n21
ahistoricity, 7–9, 30, 118, 180, 204–
205, 220, 240–42
amateurs (vs. experts), 9, 14, **17–23**,
25, 29–31, 36, **37–39**, 66, 71, 79, 90,
188, 208–14, 239, 245, 251, 296n23,
297n5, 301n10, 305n33, 325n2
Aramaic sources. *See* Criterion of
Aramaic context
archaeology, 18, 30–31, 39, 49, 93, 98,
105, 199, 213, 231–39, 250–53, 262,
300n8, 310n19, 331n36
Argument from Embarrassment. *See*
Criterion of Embarrassment
Argument from Evidence, 97–100,
103, 106
Argument from Ignorance, 188
Argument from Silence, 27, 43–44, 52,
56, 99, 114, **117–19**, 126, 188, 223

Argument to the Best Explanation, 97,
100–103, 106
Attis cult, 136, 139, 180
authorial intent, 18, 27, 132–34,
317n67. *See also* Criterion of
Embarrassment; literary emulation
axioms (of historical method), **20–37**, 39,
62, 82, 118, 191, 212, 235, 268, 269

background knowledge vs. evidence
(demarcation), 79–81, 108–109,
239–42, 276–80, 301n10, 325n5,
328n22. *See also* iteration, method of
baggage fallacy, 34–37
baptism, 142, 145–48
Bayes's Theorem: difficulties with,
60–67, 79–93, 207–82; explanation of, 49–54, 67–76; flow chart,
286–89; formulas, 50, 69, **283–85**,
329n27; for three or more hypotheses, 69, 284; historical methods
and, 97–119; odds form, 64, 69, 76,
284–85, 328–29n25; without math,
67, 286–89. *See also* consequent
probability; criteria; iteration; posterior probability; prior probability;
reference class
bias, 7–8, 61, 98–99, 114–15, 135,
227, 247, 297, 307n4, 330n28,
330n30. *See also* Criterion of
Embarrassment

333